Contents

Part IV Institutional Context

Part V Values, Identity and Knowledge

Foreword

This book on Global Environment Policies is a breath of fresh air for someone who has been involved in environmental matters for the past thirty years. I am a microbiologist, and I looked at the issues, which we tackled through United Nations Environmental Programme in a series of conventions and protocols, from the point of view of a purely scientific person. The issues of the Mediterranean and other regional seas and the Zambezi and other international fresh water resources, as well as ozone depletion, loss of biodiversity, movement of hazardous wastes and climate change were all addressed on the basis, essentially, of scientific findings. Even when I wrote, with Iwona Rummel-Bulska, our book on global environmental negotiations, I introduced every chapter with a discussion of the status of science in the subject addressed.

The book, for which I am happy to write a foreword, is different. It has two features. First, it is written by social scientists from different disciplines, although many belong to the field of political science. And, second, a majority of the authors come from the younger generation. Yet, each one of them shows an amazing depth of analysis in the subject he or she is addressing. The approach to global environmental problems is very different from all current considerations of the very sensitive issue of governance. The review of the analysis of the effectiveness of environmental agreements adds truly new dimensions different from the norm. Scientists usually adhere to an analysis of whether the agreement reversed the negative trend seen in the environment medium discussed. When I read the chapter on scientific uncertainties from the point of view of a social scientist, I see a much wider picture than what myself and other scientists are used to. I have been equally impressed by the rest of the chapters.

This volume analyses new approaches to environmental negotiations – processes of negotiation, its dynamics and effectiveness, the role of science, and the ways to influence the response of states. It also covers the issue of financing the implementation of agreed environmental treaties by less developed countries, which is so crucial to ensure proper implementation.

I am sure anyone who is interested in the issues of global environmental policy will enjoy the wealth of information presented in this book .

Mostafa Kamal Tolba
Former Executive Director
United Nations Environment Programme

Foreword

The salient environmental challenges of the 21st century derive from the very fundamentals of the 20th century. As we enter a new millennium, we have recognised that the activities, strategies, and behaviours that have characterised successful industrialisation in many parts of the global community – and the extension of the span of human life almost everywhere – has set in motion some major threats to resilience of ecological balances on which all life on earth depends.

If there is an evolving consensus, it is that the sanctity of the natural environment is seriously undermined by the nature, volume, and distribution of human actions. The scholarly community has spent the better part of the last decade reaching some broad agreement about the nature of such threat. The policy community has begun to grapple with strategies for response. And the business community has, to a limited extent, become party to these discussions. But it is a fragile consensus, bounded by imperatives and disjunctions.

Imperatives and Disjunctions

The sense that 'something must be done' is not generally disputed too much. What remains highly contentious is *what* and *how* as well as *when*. The realisation that 'no one state in the international community can unilaterally affect global outcomes' is also not in dispute. What persists in the domain of the debatable is not only the type of interventions, to be sure, but also the modality and the organisational mechanisms by which states (in the plural) can and will respond to environmental problems (in the plural and in their complexity).

Global Environmental Policies takes on a daunting task, namely to address *Institutions and Procedures*. For the first time in human history we are addressing head-on the disconnects between the challenges before us, and the tools behind us – between the nature of the problems we now face and the limitations of the conceptual frameworks we have inherited from the 19th century as these have dominated intellectual and policy circles in the 20th century.

These disconnects are best captured along two dimensions. One pertains to the disjunction between the source-locus of environmental-related problems both figuratively and jurisdictionally versus the response-locus in terms of formal decision for international resolution and global accords. The other pertains to the reality of tight coupling between social system and ecological system versus the conventional beliefs and practices that address each as a distinct, autonomous, domain of life.

Now there is little choice but to revisit the conventional views – in both scholarly and policy domains – and reconsider the very fabric of policy responses to environmental dislocations due to human activities and products of social organisations. In other words, there is 'the need for finding an integrative framework to understand emergent trends and practices of environmental protection' (p. 30) and to render such framework resilient to new understandings shaped by new knowledge of scientific, technical, economic, and social form.

Structural Conditions and Global Realities

The reality of global politics – and its legal underpinnings – is that only states are enfranchised to act on behalf of individuals. Only states represent populations; only governments represent constituencies. Individuals per se (thee and me) have no formal legal standing, nor do any jurisdictions other than the state. While there is evidence of some movement toward recognition of alternative 'voices' in international forums, only the state is recognised as 'voter'. Therein lies the heart of the dilemma.

The state consists of, and represents, a range of populations and constituencies within its borders. But the impacts of their activities transcend their borders. However, since effective distribution of political power is not 'mappable' across ecological landscapes, both factors – power and ecology – must be merged in some way to enable effective responses. Nature does not vote. Nature does not 'decide'. But Nature is not inert; it 'acts' and 'does'. And sometimes what results can be challenge prevailing social organisation in compelling ways.

This book is about institutional responses to the prevailing disconnects between the emergent ecological and environmental dislocations and the mechanisms already in place for dealing with inter-state interactions. For it is about the challenges of establishing re-connects, and creating new-connects between global political institutional and legal arrangements, on the one hand, and the diffusion of environmental dislocations due to human activities, on the other.

At issue is not which is more powerful, humans or nature – a trite observation indeed – but how to deploy the creativity of social organisation and institutional performance to facilitate mutual balance and resilience in interactions of social and natural systems.

Global Environmental Governance

Global environmental governance becomes an imperative when the problem is recognised, when it is pervasive, and when it eludes unilateral solution and, above all, when 'no action' inevitably means a worsening outcome. These properties are well illustrated in the environment domain, and can often be characterised in 'sound bites'.

Below are some of the more pervasive features of the environmental realities at hand. Such realities are characterised by uncertainty, irreversibility, and complexity. And the clichés – listed here in sound bite – aptly highlight the compelling logic for global environmental governance:[1]

- *Nature is a player.* This means that environmental effects of human action may take on unanticipated forms, whose uncertainties are sufficiently great as to insert a random element of strong proportions that cannot be contained by human action.
- *Damage is due to legitimate action.* Far from reflecting pathology and deviance, environmental damage is often due to the most normal, routinised, and legitimate behaviour, whose very nature may be condoned if not lauded worldwide.
- *Force cannot work.* In such contingencies, the deployment of troops, the most conventional instrument of force, is a singularly ineffective, if not a remarkably useless course of action, in that the response is irrelevant to the nature of the challenge.
- *Compliance is imperative.* The pervasiveness of environmental dislocation means that no one can be immune from 'attack' so to speak and that everyone's security is contingent on compliance by everyone else.
- *Doing damage by doing nothing.* Simply by choosing not to take a stand, nations can accentuate prevailing environmental problems; thus, the costs of not participating in evolving environmental accords will be equivalent to overt opposition.

Once the collaboration imperatives are recognised, when and how does institutionalisation become essential? Why are routinised procedures required?

Institutional Necessity

Of the many conditions that necessitate global institutional responses, two are commonly believed to reflect the cluster of reasons driving the quest for environmental protection. Countries collaborate (a) in the pursuit of common interests, or (b) in the management of common aversions.

In the first instance, states seek collaboration in order to, jointly, pursue some objective that they might not be able attain individually. In the second instance, the quest for collaboration is driven by the awareness that they face common adverse conditions that require co-ordinated action for effective management. This general logic presumes that countries can identify their specific preferences and objectives, as well as vulnerabilities and sensitivities. It also presumes that countries are able to identify the conditions under which unilateral action is not appropriate or bilateral operations will not suffice.

By definition, global environmental management involves self-imposed internal or external constraints on national sovereignty. Internal constraints mean refraining from taking actions that have national consequences. External constraints mean refraining from generating adverse effects outside territorial boundaries. In the environmental domain, both sets of constraints are evoked.[2] And any effective action is contingent on some minimum degree of shared understanding of the challenges at issue – knowledge of systems, events, conditions, and projected anticipation of outcomes.

Essential Triangulation

This book addresses the essential triangulation pertaining to norms and practices, issues in negotiations, and institutional contexts. By triangulation is meant the determination of the policy spaces bounded by the parameters of these three 'points'. At issue are not only the content and characteristic features of each, but also the degree of synergy – the congruence or consistency – among them. The greater the congruence, the more likely it is that institutional responses to global environmental challenges will be effective. The greater the disjunction, the more likely it is that state preferences will dominate over global priorities. In the best of all possible worlds, we would expect convergence in this triangulation. But few worlds are perfect, least of all ours. The strategic challenges therefore is to deploy the critical intellectual resources in the scholarly domain to help reduce the salience of disjunction in the policy domain – at all levels from local to global.

This book is a major step in that direction. By pointing to 'a broad spectrum of environmental policy making and institutional practice geared toward the formation of new values and international arrangements'(p. 30), it makes important contributions to the infrastructures of knowledge necessary for global governance.

Nazli Choucri
Professor of International
Relations
The MIT

NOTES

1. These bullets are based on Nazli Choucri, 'Mega Cities and Global Accords' in F. Moavenzadeh ed. *Sustainable Development in Mega Cities* (forthcoming).

2. See Peter Haas with Jan Sungren, 'Evolving International Environmental Law: Changing Practices of National Sovereignty', in N. Choucri ed. *Global Accord: Environmental Challenges and International Responses*. Cambridge, Mass.: MIT Press, 1993, pp. 401-430.

Preface

In an attempt to mitigate environmental degradation, many bilateral and multilateral agreements have been reached in the latter half of the past century. In addition, regional and international institutions and regimes were created to assess environmental situations, establish standards and support compliance with such means as technical and financial assistance. However, there has not been much systematic analysis to reveal how these efforts are effective in changing government policies and human behaviour. This book results from an attempt to fill this gap by examining global policy making and implementation in terms of values, norms, rules, and institutional procedures.

The dynamics of global politics influence the behaviour of actors in various ways. Despite regulatory activities set up in different policy arenas, the nature of global environmental management is based on voluntary agreement between state actors reached at multilateral fora. The main thrust of this book is thus how to protect the environment in a global order that does not have an overarching authority.

More specifically, the volume sheds light on the formation of guidelines and principles as well as factors that affect the effectiveness of institutional procedures set up to monitor and implement specific agreements. New organisational forms can be established to support agenda setting and policy formulation. Institutional transformation can be linked to the legitimacy of policy arrangements that enhance environmental values and norms. For instance, adjustment of the World Bank's policy making procedures in some areas has been adopted, in part, to respond to the criticism that the organisation lacks environmental concerns.

Norms and values are crucial in influencing a negotiation process as well as promoting particular policy practice. The existence of conflicting sets of belief systems and views can make environmental politics complicated. In discussion and adoption of regulatory rules, knowledge and politics are not separate domains, as the effectiveness of international agreements has to be interpreted not only in institutional but also in social and political contexts. Discussion about issues and the construction of an agenda reflects diverse economic interests and political relations between various actors.

Globalisation along with regional integration offers certain opportunities for change in the popular perception of environmental problems. The globalisation process provides a social context for the links between activities in civil society and international institutions. Given the need for involvement of the public in policy formulation and implementation, the issues of environmental governance include the role of NGOs as well as the practice of multilateral policy making agencies. Institutional changes can be connected to an organisational learning process by various actors. With the emergence of a global civil society, the network of information and knowledge bases plays a more important role in the dissemination of new values and organisational learning for institutional change.

The contributors represent broad intellectual traditions in international relations and other fields of social science. Each chapter has a strong conceptual framework that is applicable to specific aspects of global environmental politics. In particular, theories on international organisations, regimes, political economy and decision making have been utilised to investigate various sets of policy questions relevant to institutional changes, agenda setting and the evaluation of implementation strategies. On the other hand, a general sketch of grand theories on global governance is avoided in order to focus on particular practices in diverse policy making realms. Overall, the strength of the book lies in linking policy making and implementation to knowledge, value, and structural transformation.

This volume will appeal widely to policy makers as well as academics with a special interest in environmental issues. Practitioners will benefit from the analysis of negotiation and policy implementation processes. The insight for future action can be provided by the investigation of the political, economic, and social contexts of value formation involved in shaping the global environmental agendas. In considering its thorough examination of international responses to environmental problems, the volume can also be adopted as a textbook for many courses at upper level undergraduate and postgraduate courses.

Completing the book project has required coordination among the authors and serious discussion with other experts. I appreciate, first of all, the contributors who patiently responded to the request for revisions. I am also very grateful to Dimitri Stevis, Paul Wapner, Marty Rochester and others who either reviewed the chapters or lent their insight through consultation on various subject matters covered in the book. Gloria Rhodes patiently formatted the manuscripts and provided research assistance. Vivian Leven and Adina Friedman corrected various types of errors in the manuscript. Finally, I also wish to thank Jim Whitman, the general editor of Palgrave's Globalisation Series, for his support and encouragement. In finishing this book, I am thinking of my parents, Mary and our small daughter Nimmy who devotes her love and attention to this planet. This book is dedicated to the future generation who must embrace the journey toward restoring a peaceful relationship between the natural world and all its inhabitants.

May 2000
Ho-Won Jeong

Notes on the Contributors

Charlotte Bretherton is Lecturer at Liverpool John Moores University. Her articles on ecofeminism and environmental politics have been published in various journals of international relations and environmental studies.

Pamela S. Chasek edits the *Earth Negotiations Bulletin*, a reporting service on environment and development negotiations within the United Nations system. Dr. Chasek is an adjunct professor at Columbia University's School of International and Public Affairs and Manhattan College. Her publications include: *The Global Environment in the 21st Century: Prospects for International Cooperation* (1999), *Global Environmental Politics*, 3rd ed. (with Gareth Porter and Janet Welsh Brown, 2000), and *Managing Global Resources: Lessons Learned from 25 Years of Multilateral Environmental Negotiations* (2000).

Elizabeth R. DeSombre is Associate Professor of Political Science and Frost Associate Professor of Environmental Studies at Wellesley College. During academic year 1998-9, she was a visiting scholar at the Weatherhead Center for International Affairs and Harvard University. Her research has focused on the process of encouraging international environmental cooperation, and on the use of economic sanctions and economic aid for achieving environmental goals.

Gabriela Kütting is Lecturer at Department of Politics and International Relations, University of Aberdeen. In addition to her recent book on the effectiveness of international environmental regimes, she has authored articles on environmental changes and policy responses.

Ho-Won Jeong is on the faculty of the Institute for Conflict Analysis and Resolution, George Mason University. His current research interests include environmental policy making and identity politics. He has published several books and numerous journal articles in the field of conflict resolution, development and environmental issues. His most recent book is *Peace and Conflict Studies: An Introduction* (2000).

Ronnie D. Lipschutz is Associate Professor of Politics and Associate Director of the Center for Global, International and Regional Studies at the University of California, Santa Cruz. He is the co-editor (with Ken Conca) of *The State and Social Power in Global Environmental Politics* (1993) and author of (with Judith Mayer), *Global Civil Society and Global Environmental Governance* (1996). His most recent publication is a volume co-edited with Beverly Crawford, *The Myth of 'Ethnic Conflict': Politics, Economics and 'Cultural' Violence* (1998).

Rodger A. Payne is Associate Professor of Political Science at the University of Louisville. Most of his latest research focuses on transnational relations, global environmental politics and the democratisation of development. His articles have appeared in *Journal of Democracy, Journal of Peace Research, Policy Studies Review*, and various other journals. Currently, he is working on a book on transnational environmental politics.

Marvin S. Soroos is professor of political science and public administration at North Carolina State University, where he has taught about global problems and policies, with emphasis on the environment, since he joined the faculty in 1970. His books include *Beyond Sovereignty: the Challenge of Global Policy* (1986) and *The Endangered Atmosphere: Preserving a Global Commons* (1997). He is also co-edited *The Global Predicament: Ecological Perspectives on World Order* (1979). His articles have appeared in the *International Studies Quarterly, International Organization, Environmental Review, Policy Studies Journal, Environment*, and *Human Ecology*.

Part I: Introduction

1 Politics for Global Environmental Governance

Ho-Won Jeong

The environmental challenges brought about since the Industrial Revolution are unprecedented in human history. This is astonishing especially considering that population growth and consumption curves were almost flat for the previous hundreds of millennia (Choucri, 1993, pp. 8-9). In an attempt to reverse further deterioration in the global environment, scientific research has been conducted, and numerous negotiations have been held to reach agreements to regulate environmentally harmful activities. Social consciousness of problems has generated environmental movements both at local and global levels. Owing to the ineffectiveness of the existing mechanisms in reversing the current level and speed of ecological destruction, however, the issues of global environmental governance continue to deserve full attention in 21st century global politics.

The complexity of relationships between society and nature resists any easy solutions to environmental problems even at a scientific level. Social structures and processes influence our perceptions of the natural world, while new values and knowledge create the necessity of rethinking relations between human activities and environmental changes. Setting up priorities in the politics of global environmental management is entangled with not only conflicting interests but also political and scientific uncertainties. Diverse perspectives on the causality of problems are related to the complexities of issues and their linkages. A lack of trust in voluntary compliance with the international agreements is added to difficulties in controlling the behaviour of a large number of social actors who affect environmental conditions.

3

One of the main issues continues to be how to restructure existing institutions in order to alter practices that cause negative effects on the environment. Multilateral cooperation has been inadequate to cope with global environmental challenges, and new forms of governance need to emerge from changes in policy-making practice. Given the nature of environmental politics and the magnitude of the issues, states cannot be conceived of as the single dominant actors (Soroos, 1999). Global interdependence requires limits on political sovereignty and the state's pursuit of its own self-interests. There have been numerous calls for the increasing participation of sub-national groups and transnational organisations in value formation and global policy co-ordination.

Global environmental governance has to focus on an array of functional issues, ranging from climate change, ozone depletion and the trans-boundary spillover of pollutants to loss of biological diversity. Governance, especially in an environmental policy making setting, can be understood in sociological terms. Because interactions do not take place in a social vacuum, it is important to understand the constitutive principles of various relationships. If the goal of governance is considered from a collective problem-solving perspective, attention needs to be paid to the process of setting up rules of conduct that define acceptable practice and guide interactions between self-motivated social actors. At the same time, it needs to be recognised that the existing rules can be challenged and transformed by the emergence of new identities and norms. How knowledge and values influence consensus building for global action is an important empirical question in investigating the effectiveness of environmental policy formulation and implementation.

In recognition of the fact that the earth is an interlocked single ecosystem, this book highlights social and political processes to promote common understanding, generate new values and enhance consensus in tackling global environmental problems. The mediating role of social institutions is crucial in redefining the values and norms of both state and non-state actors as well as co-ordinating their behaviour. In responding to the above theme, the contributors examine regime formation, global civil society, ecological identity, international negotiations, effectiveness of agreements, international organisations and other aspects of environmental policy making and implementation.

GLOBALISATION CONTEXT

In an environmental context, globalisation reflects both the prehistoric and historic tendency of the human species to expand their activities. The effects of globalisation are complex: economic integration and expansion all have a significant impact on environmental change. The environmental consequences of globalisation can only be assessed in a long-term frame (OECD, 1997, p. 8). The globalisation of the planet, which has proceeded for millennia, is based on 'the intense interactivity and interdependence of human population growth, technological advancement, and pursuit of resources'. Thus, globalisation 'emerges as a logical and compelling anthropogenic process' (Choucri, 1993, p. 8).

Ecological degradation encompasses global dimensions beyond cultural differences, divided political jurisdictions and diverse economic systems. In the modern era, the nature of environmental issues in different parts of the world becomes similar due to global economic integration resulting from free movements of capital and goods associated with the process of industrialisation. Consumerism has been encouraged by the culture of modernity and the expansion of a market economy.

Most environmental problems have a universal impact on people from every corner of the planet, in contrast with armed conflicts and human rights violations that may be visible only in specific societies. Greenhouse effects and the general depletion of the ozone layer are globally felt, while the large proportion of their main causes can be attributed to a few countries. Complex ecosystems with shared river and lake basins transcend national boundaries. Desertification as well as air and water pollution can be intensified by geophysical linkages. Ecological degradation on a global scale reflects the cumulative effects of many changes in local material conditions. As problems are interrelated, one country or continent is not immune from the environmental consequences of a major disaster. A global identity can be assembled by the fear of threats to the global environment (Yearley, 1996).

Environmental conflict has been globalised as well as regionalised, and its nature reflects an asymmetric relationship between victims and polluters. Activities such as excessive burning of fossil fuels occurring within one state may contribute to national economic growth but negatively affect the welfare of people in other national jurisdictions. The destruction of tropical rain forests for farming and extraction of minerals is motivated by regional economic interests but its effect is global via climate changes. Deforestation in the Himalayas has global ecological ramifications beyond catastrophic flooding in Bangladesh. Other examples include the destruction of a whale sanctuary in Mexico due to a Japanese multinational corporation's industrial projects. Ecological globalisation in many guises presents difficult problems not easily resolved in a traditional governance structure (French, 2000, p. 185).

ECOLOGICAL SYSTEMS AND SOCIAL ACTORS

The earth is viewed as one system, and activities of each part are related to the whole. As suggested in the Gaia hypothesis, the interplay of all the planet's elements produces an outcome greater than the sum of its parts. Ecological, geophysical, geochemical and biogenic processes are affected by a multiplicity of interacting factors that operate together to form a natural system. Thus, ecosystems can be characterised by the total complex of relationships involving the exchange among the participating organisms and their physical environment (Milton, 1996, p. 56). As changes in one area lead to changes in others, various problems intersect with each other in unpredictable ways.

From the perspective of a holistic ecosystem, activities at different levels of social and natural systems can be analysed to examine the interactive nature of various species. The well-being of living species on the planet is influenced by the outcomes of decision making at all levels of social aggregation and responses in all parts of the natural environment. Feedback between human and non-human actors has both social and environmental implications and is not a mere reflection of biological and physical phenomena. Therefore, the conceptual understanding has to focus on the relationships of humans with their natural surroundings.

The environment is transformed by the way various organisational forms operate, including family, neighbourhood, community and large corporations. Social structures from tribal bands to modern states 'have made it possible for individuals to manage their activities in order to obtain what they demanded from the natural environment' (Choucri, 1993, p. 11). Individuals or groups make demands and act upon natural and social environments in order to satisfy their felt needs, wants, and desires. Economic systems have been (re)produced by the practices of organisations that apply knowledge, skill and technology to nature. Industrial interests are served by transforming raw materials into economic wealth.

The decisions of social organisations and other major collectivities have an impact on resource expansion and depletion, and distribution between and within their boundaries. Relations between humans and nature are influenced by changes in the technological, economic, socio-cultural, political and other capacities of organisations. Localised methods of production are different from the massive production capacities of international capital. For instance, genetic diversity is crucial for the survival of indigenous populations who depend on other species for food and medicine. Indigenous peoples have not been tightly integrated into an international market, and they are socially and culturally adapted to the local environment for their subsistence (National Research Council, 1992, p. 140). Biological and physical resources in indigenous land are seen as objects to be exploited for profit by industrialists, and the survival of the local ecosystem has been threatened by the accumulation of wealth and other state building projects of modern political elites (Kamieniecki and Granzeier, 1998, p. 262).

Increased pressure on nature by the expansion of human activities has destabilised the feedback relationship between social systems and nature. In the industrial age, 'it became increasingly evident that exponential population growth and rapid technological development, in combination, were overburdening the natural environment and creating policy dilemmas' (Choucri, 1993, p. 13). Intensive economic activities have proceeded without much consideration of the sustainability of natural resources and changes in sources of water and energy systems. It is often argued that the excessive demand for more resources can be overcome by changes in technology. However, technological advances would not be able to compensate for the continuation of rising consumption levels combined with rapid population growth. Natural resources do not increase dramatically in response to a growing demand for a higher standard of living.

The arrival of an industrial culture with a global scale magnified the human impact on nature in force, intensity and ubiquity (Callicott, 1994, p. 10). Environmental change extends over long periods of time with far reaching consequences in many areas. For example, the destruction of both animal and plant species in the natural habitat will diminish the social and physical aspects of life of the future generations as well as the present. In many poor parts of the world, economic disparities and deteriorating living conditions are attributed to the mismanagement of resources. The prospect of the collapse of stable relations between the environment and humanity caused by trends of exponential growth has been indicated by several scientific models since the late 1960s (Meadows, et al., 1972). Despite disagreement over the extent to which current economic activities contribute to pollution and resource depletion, it is generally acknowledged that the biosphere does not return to equilibrium once it reaches a certain level of stress.

Given that all life is dependent on the planet's land, water and air, policy debates about environmental issues have been put in the context of how to foster sustainable human economic activities. The poor quality of the environment has costly effects on agriculture, employment and other aspects of human well being. For instance, human society is directly affected by health hazards attributed to air and water pollution. Such changes in the ecosystem would require adjustment of social systems because of the impact of the disruption of the natural world on human activities and relations between societies.

Exploring solutions to global environmental problems requires understanding a complex interweaving of political processes, economic development strategies and social values at different levels. Explanation about the causes of greenhouse effects can be scientifically offered, but solutions have to be found at social and political levels. The transition to solar, wind, hydro, and biomass energy sources, away from coal and oil, would not be successful through the development of new technologies alone. Such anti-greenhouse measures as prevention of deforestation and shifting from fossil fuel use will inescapably bring about changes in nearly all social and economic activities that produce carbon dioxide (CO^2).

While technological advances help resources to be used more efficiently, overall solutions would not be found without dealing with the unique capacity of the human species to destroy the habitat shared with other species. Polluters have less motivation to control the sources of pollution given its spread impact among multiple actors. Regulations as well as subsidies and other incentives have been suggested to control behaviour and policies of economic actors. Corporate investment in research on substitutes for chlorofluorocarbons (CFCs) resulted from the 1985 Vienna Convention for the Protection of the Ozone Layer and the 1987 Montreal Protocol on Substances That Deplete the Ozone Layer that require phasing out CFCs as a step towards repairing the ozone hole.

However, changes in values and attitudes related to resource-using behaviour require more than economic incentives and regulations. Social variables reveal motivations and actions needed to reverse the current trend. Such issues as ecological sustainability and societal vulnerability have been made an unavoidable part of public discourse. In adjusting their actions to new environmental realities, social actors have different abilities to learn from experience. Social responses to environmental change play an important role in shaping the nature of environmental politics.

The management of the environment is inevitably influenced by an increase in the complexity and specialisation of social organisations that have diverse knowledge bases and belief systems. Given their unique experiences vis-à-vis nature, social actors have different perceptions and interpretations of favourable and unfavourable environmental consequences. Growing scientific knowledge and public understanding do not necessarily translate into shared responses to the problems (Conca, et al., 1995, p. 10). Changes in social practices over time would result more from a prolonged process of public discourse than short-term policy responses aimed at immediate effects on behaviour. It needs to be noted that many limits on the ability of human beings to manage a complex system will slow structural transformation. The continuing concern would be not only how to co-ordinate the activities of various groups and organisations in formulating global policy responses but also how to overcome a traditional structure of values that permeates the whole system.

LIMITATIONS OF NEOLIBERAL ECONOMIC NORMS

Environmental degradation would not be eased by the present growth model that leaves increasing numbers of people poor and vulnerable. In understanding the impact of modern political economy on changes in the ecological system, it is a widely accepted fact that markets 'are typically not sensitive to local ecological conditions' (OECD, 1997, p. 9). Many large-scale development projects erode sustainable economic bases for indigenous populations, as well illustrated in the Palonoroeste Project in Brazil (funded by the World Bank) that led to destruction of rainforests in the Amazon. Environmental concerns are not given priority in many Third World countries that not only face such a pressing need for survival as food, energy, etc. but also struggle to achieve financial and monetary stability. For instance, logging rainforests has been used to look for an exportable means of exchange by debt-stricken Third World governments.

In traditional approaches to environmental management, free market solutions are advocated for efficient use and allocation of resources. At the same time, growth orientation remains a dominant norm both at national and international policy-making levels. The sustainable development model suggested by the Brundtland Report (World Commission on Environment and Development, 1987) is still framed in the traditional interpretation of the relationship between the environment and economic growth. It asserts that growth should be the main path to and most effective means for reducing poverty, which is a main cause of environmental degradation. Thus the Brundtland Report does not directly challenge industrialism and consumerism, nor does it seriously question their undesirable consequences for the environment. Since the early 1990s, its notion of sustainable development has been integrated into the neoliberalism of the World Bank. However, adjustment and reforms within neoliberal economic norms leave intact the basic socio-economic forms considered to be detrimental to restoring ecological balance.

The contradictions between ecological sustainability and economic growth cannot be resolved within a traditional paradigm that resources are converted to create monetary value in the reaffirmation of the dominance and efficacy of capitalist industrialism. Through their role in the redistribution of resources, markets play a significant role in determining environmental values. The assumption of unlimited substitutability of environmental goods unpriced at the market place underestimates the fact that some of the important functions in the natural world, if ever lost, will not be recovered within a realistic time frame (Ayres, 1998, p. 7).

In formal economic models, production is framed in the context of effective utilisation of capital while neglecting the productive roles of unpriced environmental resources. Whereas the trade-off between cheaper goods and better environmental quality are often made in a free market economic system, social costs increase with the irresponsible use of collective resources such as air and water. As ethical and political considerations should not intervene in the functioning of a free market system, competing interests and objectives have to be mediated by market principles, often detached from social realities.

Contrary to classic economic theories, the mutual adjustment of supply and demand of resources is slow in the cycles of recession and expansion of the market (Meadows, 1991, p. 244). In a continuing growth model, it is not seriously considered that the process of consumption and production of commodities is unstable in the long run. The effects of industrialisation and growth on an ecological system are not neutral and cannot be easily controlled within an unregulated free market system (Rees, 1992). Environmental crisis is inevitable if every developing country seeks the same level of consumption enjoyed in the West while people in advanced industrialised countries are reluctant to change their life styles.

The profit motive in a free market economic system does not count ecological values since monetary calculations determine the choice of the means for production (Carley and Christie, 1993, p. 56). As the market economy is driven by the profit-oriented objectives of private firms, capital and labour are strategically deployed to reduce the cost of production. Incentives for production come from the ability to increase market shares, and reduction in the input cost is directly related to success in a free market. The main concern of enhancing efficiency and productivity results in neglect of the social and ecological costs of polluting water and air. Although the market is seen as a remarkably efficient device by conventional economic theories, 'it does not respect the sustainable yield thresholds of natural systems' (Brown, 2000, p. 9). With economic interests overriding ecological concerns, it is often overlooked that short-term economic gains are achieved at the expense of long-term environmental stability.

Recently there have been some efforts to expand the existing system of national accounts to environmental and social dimensions. In addition, such instruments as taxes on emission can be introduced to encourage less consumption of polluting productions and the development of cleaner technology (Pearce, et al., 1990). Yet incorporating environmental resources in an accounting system and designing market based measures alone would not solve fundamental problems arising from consumption patterns and uneven distribution of resources. In a neoliberal environmental logic, the efficiency of market mechanisms is still considered as a universal mechanism rather than being 'evaluated against the goals it must achieve' (Stevis, 2000, p. 74). Radical changes in our life styles have been called for in order to curb the negative impact of the current trends (Smith, 1992). Ultimate solutions to environmental problems will have to be anchored in the transformation of an economic system that, in part, is made possible by changes in value systems.

ENVIRONMENTAL VALUES

The way boundaries between humans and the environment are conceptualised and constructed reflects value issues. Policy differences are based on different conceptions of the role of humans in changing the environment. In anthropogenic views, the value of environmental objects lies in their usefulness to humans. Nature is manipulated, conquered, and changed to provide material bases for industrial, agricultural and other economic activities. The anthropogenic attitudes and patterns of thought represent a traditional Western society's assumption that nature exists to serve humans and to be exploited for human benefit.

Humans assume power over the material world with the rise of modern scientific research and technology that provides an ideology of progress (Carley and Christie, 1993, p. 70). Nature is reconstituted as a system 'that can be dismantled, redesigned and assembled anew to produce "resources" efficiently and in adequate amounts when and where needed in the modern market-place without seeing a degradation in carrying capacity' (Luke, 1995, p. 28). The treatment of nature as something separate from its own intrinsic value culminated in Enlightenment thinking. As change in the means of knowing the world is based on the development of scientific methods, modern science does not lend nature its own value independent of human beings. The material and social prosperity of industrial society depends on the subjugation of nature through scientific knowledge and technology. Because nature is not regarded as an end but as a means, knowledge acquired by scientific analysis could be put to work to give the human race mastery over nature.

The materialism of industrial culture undermines ecological ethics with the destruction of the image of the earth as a living organism (Callicott, 1994, p. 10). Nature exists completely outside of the moral sphere for industrial developmentalism that considers the application of technology to nature as improvement (Merchant, 1996, p. 78). The domination and mastery of nature have become core concepts of the modern world with the increasing mechanisation of Western culture along with the continuation of commercialisation and industrialisation.

The environment is seen, without cultural constraints restricting the actions of human beings, as an instrumental value for human beings. In managerial approaches that put the superiority of humans on other components of the planet's biophysical system, mostly technological solutions are sought. An overemphasis on technological solutions encouraged by the instrumental view neglects the intrinsic value of wildlife and landscapes (Carley and Christie, 1993, p. 78). Technocentric management is compatible with a faith in the capacity of technology to harness nature (Rogers, 1994). It mostly aims at reducing the extent and speed of exploitation and pollution of nature that has been pursued to the extent to ignore a potential for the collapse of the entire ecosystem. Its primary goal has been to protect human health and a few species of special interest along with conservation of resources.

With the recognition of the intrinsic value of environmental objects and their rights to co-existence with humans, non-anthropocentric views perceive human behaviour to be mostly responsible for environmental crisis. The dominant culture of society and existing political order engender materialism with threats to life on the planet. 'Ecological values embody an appreciation of nature in all its varieties ... Humans are seen as but one species among many, but also as the one species whose dominance could be so thoroughly going as to threaten many, if not most, other species and thereby itself' (Paehlke, 1995, p. 132).

In considering that humans are an integral part of nature, an ethical argument can be made that they have no right to distort the natural environment (Foster, 1997). Communitarians advocate resource preservation with limited growth and control of greed. Sustainability can be achieved through structural reform encouraged by the decentralised initiatives of social movements. Changes in underlying institutional structures, social values, government policies and economic functions are a prerequisite in a holistic approach to establishing a new relationship between humankind and their environment.

Radical ecocentric movements oppose established institutions and reject compromise in favour of a thorough transformation of advanced industrialised society. In bioethicist views, overcoming human centredness is based on living in harmony with nature. By drawing on Western anarchist traditions, social ecology attributes ecological problems to human social hierarchy. On the other hand, ecofeminism emerged in the 1970s to link the domination over women to the exploitation of nature (Plumwood, 1993).

Ecological egalitarianism is represented in the principles of diversity and symbiosis with a deep-seated respect for the ways and forms of life. In deep ecology, nature is seen as 'the ultimate source and location of values' (Smith, 1999, p. 374). Thus humans have to adapt themselves to nature based on the recognition of their equal value to other species. Organisms are considered as knots in the biospherical net or field of intrinsic relations. Overall, ecocentric views seek radical social transformation in favour of building self-sufficient, decentralised communities that respect the intrinsic value in all of nature.

INSTITUTIONS AND REGIMES

Institutional arrangements for governance arise from the situation that one actor's action impinges on others' welfare. With the pursuit of competing goals, activities interfere with and impede each other. Governance 'involves the establishment and operation of social institutions' for alleviating collective action problems created by joint losses attributed to incapacity to cooperate (Young, 1994, p. 15; Young, 1997, p. 4). It does not presuppose a central administration and enforcement structure (Caldwell, 1996).

Interactive decision making derives from a high level of interdependence by which the well-being of one actor is affected by the actions of other actors (Young, 1997, p.3). Cooperation is necessary to control the unregulated use of the environment as the public good. In an interdependent situation where individual actions are ineffective without co-ordination of actions with others, institutions play an important role in determining regularised patterns of interaction and collective behaviour. The emergence of a new set of problems often invites the conditions for the creation of new institutions or changes in the role of existing institutions.

When multiple users have access to the resource domain for their individual gains, its abuse hampers common welfare with undesirable outcomes to everyone. In mixed motive situations, institutions can create incentives for cooperation by promoting mutual gains. The preferences of action are affected by redefining the identities and interests of actors. Deepening and broadening interdependencies have been intensified by technological and economic changes in various functional areas such as the environment and trade. The role of institutions is crucial in building political capacity for global environmental governance.

Institutions, serving as 'social artifacts' that fulfil the need for co-ordination of activities, can change the collective outcome with sets of rules of the game or codes of conduct. Thus, institutions come into existence with not only the consciousness of common interests and values but also the desire for a common set of rules which bind social practices and guide interactions among states. Arrangements can be flexibly adjusted to new situations, or decay under internal or external pressure. Institutions have varying degrees of formalisation, authority structure and functional scopes, and unlike organisations they do not need to be represented by material entities.

Regime theory perspectives have been broadly applied to an attempt to find solutions to problems with global environmental policy co-ordination within a multilateral institutionalised framework. The concept of regime has been used to describe 'the institutional arrangements established under the terms of treaties and conventions and the established practices that have evolved more informally in international society' (Young, 1994, p. 205). As institutional responses to a collective situation, regimes are supposed to regulate the behaviour of the participants in the process of defining social practices. Institutions that govern the interactions of actors can be formed in the process of regime creation.

In general, regimes are seen as a set of norms, rules and decision making procedures producing common expectations in a particular issue area. Whereas rules and norms of regimes can be created by treaties, legally binding formal arrangements may not be the only means of regime formation. Since a significant level of common understanding and consensus is a prerequisite to joint actions, the constitutive provisions of an institutional arrangement have to be accepted by actors. In addition to encompassing the explicit rule systems structuring international decision making, environmental regimes can recognise implicit principles and norms in changing actor behaviour (Mason, 1999, p. 215).

In neo-institutional approaches, international regimes offer self-help strategies in a structure of regulated anarchy by helping overcome obstacles to co-ordinated actions. Formal organisations and their structures may be set up to promote a set of rules and prescriptions for representing collective choices. The impact on participant behaviour and practice can be made through the incentive structure of a regime (Stokke, 1997, p. 42). Whereas a regime is portrayed as a rational institutional response to collective suboptimality problems, the rationalist portrait of agents as self-contained units with fixed preferences can be criticised by a more sociologically inclined approach.

Policy co-ordination at a collective level can be facilitated by common recognition of the issues and agreement on the means to achieve the goals. The effectiveness of international environmental regimes can be measured by the ways in which treaties, conventions and declarations regulate behaviour of both state and non-state actors. International rules and policies are agreed upon and implemented mostly by state actors, but non-state actors participate in the maintenance of the regime through their monitoring activities and their input in the rule making process.

Formal rules and agreements have not yet been made in some serious areas while other areas, for example, ozone depletion and trade in endangered species, observed the emergence of a concrete set of agreements, expectations and goals (Nanda, 1995). The 1973 Convention on International Trade in Endangered Species of Wild Fauna and Flora (CITES) has brought order into the interactions of sovereign authorities. International Tropical Timber Agreement was reached in Geneva to provide an effective framework for the sustainable utilisation of tropical forests. The 1979 Convention on Long-Range Transboundary Air Pollution and its subsequent protocols provides systems of rights, rules and relationships.

Other environmental issue areas have also been managed by adopting norms, institutions and legal instruments requiring individual actors to give up part of their rights. The Antarctic treaties allow cooperation in scientific research while preventing the exploration of minerals under the ice. The global dumping regime based on the 1972 London Convention attempts to forbid the disposal of hazardous waste at sea. The UN Convention of the Law of the Sea was adopted in 1983 to set up a comprehensive legal regime for the sea, including protection and preservation of the marine environment.

Whereas new international norms and practice can be set up by the influence of powerful states (hegemons), prevailing collective situations such as policy interdependence over time lead to regime formation. Regimes in environmental issue areas arise from a process of converging expectations or explicit or implicit bargaining over rules rather than imposition by dominant actors. The socio-political context of regime formation is characterised by the interplay of interests, ideas, knowledge and information that serve as power sources. Epistemic communities comprised of transnational networks of experts have an impact on the public and policy makers in rule making.

Not being bound by territorial jurisdiction, regimes vary in terms of their membership, functional scope, geographical domains, and degrees of formalisation and stages of development (Porter and Brown, 1996). The regime for Antarctica comprising the several components of the Antarctic Treaty system has a regional focus with the goal of preserving a vast and fragile ecosystem. The ozone regime, consisting of the 1985 Vienna Convention and the 1987 Montreal Protocol as amended in London in 1990 and in Copenhagen in 1992, has a global scope, and its jurisdiction is not limited to specific geographical areas.

As regimes are not static and evolve with changes in perceptions and power relations in the issue domain, the terms of constitutional principles and rules can be re-negotiated through institutional bargaining with the assistance of standing international bodies. Whereas demand for changes reflects alterations in the ecosystem and emergence of new issues, regime transformation is ultimately affected by the interplay of different interests and motivations, institutional bargaining structure and the availability of new scientific knowledge.

Changes in institutions, rules and procedures affect a particular regime's functional scope, jurisdictional attributes, geographical domain or membership (Jurgielewicz, 1996). The Antarctic Treaty System (designed for the regulation of human activity and interstate relations on all lands and waters of the region) started with the efforts to preserve the continent as a non-militarised zone of peace and international scientific cooperation (the 1959 Antarctic Treaty). Since the time of its foundation in 1959, its membership grew and its functional and legal arrangements gradually expanded beyond its original mandate to promote cooperative scientific research. This helped the participants' actions be transparent and responsive to concerns of outside actors despite their operational autonomy.

Rules and decisions must have a binding effect on all members in order to bring about a desirable outcome. Expected outcomes cannot be achieved if there are free riders who do not put limits on their practice while taking advantage of others' constrained behaviour. Improvement in air quality, protection of the ozone layer, reduced global warming, and the conservation of natural resources will benefit even those who refuse to take the burden. Given that mutual compliance is in the best interests of all, convincing free riders to join the system by a combination of threats of sanctions, moral persuasion and tangible material benefits becomes an important task in regime maintenance.

Regimes prove their significance when the behaviour of states conforms to common understandings and rules agreed upon by their members. Some rules are legally binding while others depend on voluntary enforcement to guarantee compliance. Rules can be more easily enforced when a system to detect violations is created, and collective measures are taken to sanction violators and provide incentives for fulfilling obligations. Compliance mechanisms are supported by a structure of rewards and sanctions. Material rewards such as funding can be contingent upon satisfactory review of programmes and proposals. On the other hand, non-compliance may lead to denied loans and suspended technical aid.

Not only the submission of regular reports by states but also independent inspection conducted by treaty secretariats are the prerequisites for evaluation of compliance. Monitoring activities of a national government can be ineffective due to the poor design of reporting systems as well as a lack of will (Young, 1999). Transparency in compliance is enhanced by effective information systems. Peer pressure can be used to bring compliance, but its effectiveness heavily depends on the prevailing social norms. The desire and value have to be placed on maintaining a good reputation.

INTERNATIONAL ORGANISATIONS

International organisations get involved in setting agendas and monitoring implementation through data collection and information exchange. Scientific studies are supported by improved capabilities for gathering and assessing data before determining which activities are likely to affect environmental conditions. International organisations fulfil the role of providing independent sources of information and analysis of the growing environmental abuse and the associated economic and political implications (Feldman, 1995). In addition, international agencies have been conducting and publishing environmental performance reviews of their member countries.

In the co-ordination of responses to global environmental problems, the existence of different national standards and regulations creates technical difficulties for transnational co-ordination. By monitoring and circulating information about the quality of the worldwide environment, the United Nations Environment Programme (UNEP), headquartered in Kenya, contributes to consensual knowledge building. It has contributed to the development of international guidelines, recommendations and norms approved by the UN General Assembly. Their proposals draw respect though they are not legally binding (Trolldalen, 1992, p. 35). UNEP was explicitly created to cover a variety of environmental issues, ranging from control of air pollution and protection of the ozone layer to biological diversity.

In support of other agencies, UNEP not only provides information about environmental quality but also finances the protection of tropical forests, wildlife preservation and other projects. In collaboration with the World Wildlife Fund and the International Union for the Conservation of Nature and Natural Resources (IUCN), UNEP launched the World Conservation Strategy in 1980. The plan aimed at preserving genetic diversity and ensuring the sustainable utilisation of species and ecosystems (Mather and Chapman, 1995, p. 247). UNEP was given a broad co-ordinating role to oversee the work carried out by other agencies in the areas of ozone depletion (Breitmeier, 1997, p. 98). Its role also includes convening numerous ad hoc conferences on specific environmental problems, a typical example being a 1977 scientific meeting on threats to the ozone layer.

Other organisations focus on single or few specific issue areas such as global warming and the protection of endangered species. The World Meteorological Organisation (WMO) conducts scientific research and is engaged in monitoring global climate changes. The International Whaling Commission regulates commercial activities in order to prevent the elimination of whales. The International Maritime Organisation is concerned with reduction in pollution through overseeing shipping activities. The UN Commission on Sustainable Development was created in 1993 to reflect on specific concerns with the links between development and the environment following the Rio Environmental Summit. Over the last several decades, other environmental organisations that are responsible to member states have emerged from international agreements.

International organisations in specific areas can perform a central function in co-ordinating activities. The UN Food and Agricultural Organisation (FAO) plays a monitoring and advisory role on forest management. The Tropical Forest Action Plan was promoted by the FAO for the purpose of slowing down deforestation. In order to enhance the sustainable management of forests, the Plan established a forum for development agencies to co-ordinate forestry policies and help developing countries formulate their own forestry management programmes.

In collecting further scientific knowledge about the climate and global warming, the WMO has built a long history of collaborative relationships with the International Council of Scientific Unions. The implementation of the World Climate Program (WCP) adopted at the second World Climate Conference in 1990 was supported by UNEP and UNESCO. In the atmospheric policy area, the WMO marshals monitoring systems and scientific expertise whereas UNEP is better equipped with formulating and implementing international law and policy. Global and regional monitoring by the WMO's Climate Observing System enables compiling environmental pollution data. The WMO co-ordinates activities of the Global Ozone Observing System (as part of the Global Atmospheric Watch System) that measures stratospheric ozone concentrations. The programme involves more than 140 national stations that collect, exchange and analyse data.

International organisations receive reports on treaty implementation by states and facilitate independent monitoring and inspection. Administrative secretariats have been created to establish and monitor the detailed protocols of treaties and conventions. The International Tropical Timber Organisation was formed in 1983 to protect the future of timber trade and conserve the forest ecosystem. In addition to compliance monitoring, international organisations are engaged in developing statistics showing the extent of activity in the regulated field. For example, fish catch data are submitted to FAO. The data of CFC production, trade and consumption are sent to the Ozone Secretariat.

Compliance requires implementation capacities, and some international organisations provide technical support for reduction in emissions of pollution gases and control of banned practices within jurisdictions. In capacity building, the FAO and its regional fisheries commissions co-ordinated programmes for efficient fisheries management in developing countries. Technical and scientific advice is available to countries combating transboundary environmental threats to the atmosphere, biodiversity and water resources. Development and transfer of environmentally sound technologies for climate and biodiversity have been funded by the

Commission on Sustainable Development in collaboration with other UN specialised agencies. The Global Environmental Facility that became operational in 1991 is run jointly by the World Bank, the UNDP, and UNEP. It was originally designed to funnel loans to low-income developing countries whose projects need to be financed.

NONGOVERNMENTAL ACTORS AND GLOBAL CIVIL SOCIETY

The role of nongovernmental organisations (NGOs) has become increasingly visible since the emergence of ecological concerns in the late 1960s. Environmental movements, both local and global, have been pressing policy makers to recognise their concerns. NGOs range from peasant women protesting deforestation in poor countries to a worldwide network of groups campaigning for diverse issues. Local efforts to resist environmental degradation reflect the struggle of marginalised groups for survival. On the other hand, Greenpeace, Friends of the Earth and other activist and advocacy groups were created to appeal to public opinion, influence government policies, and support treaty regimes and international organisations. Large national and international NGOs can suggest new policy initiatives, monitor compliance, offer scientific advice, and lobby to promote strict international standards for environmental protection.

Non-sovereign actors enhance social and cultural interactions in a global civil society where decentralised networks of social institutions represent different interests and values (Princen, 1994). Business seeks profit while other types of organisations pursue political, social and cultural goals. In contrast with the roles of economic actors such as multinational corporations, the activities of non-governmental organisations are driven by values, and they act for felt concerns and other non-selfish reasons. Most environmental NGOs promote environmental values, but some groups affiliated to the Catholic Church and other conservative religions opposing population control hold viewpoints that conflict with sustainable environmental policy.

Many NGOs defend the rising environmental interests, often challenging governmental actors, with their qualifications for transnationalism, transparency and legitimacy (Princen, 1994, p. 34). The relevance of NGOs to global politics is found in shaping the behaviour of other powerful actors by influencing public views. By forming a coalition in specific issue areas such as the protection of rainforests, NGOs attempt to win broad social approval for their agendas. They help expand the social base for environmental movements through increased communication and exchange of information among individuals and groups around the world. Alliances are built in the process of generating and transmitting new knowledge and ideas beyond national boundaries.

In pursuing environmental problems as a political agenda, NGOs adopt a variety of strategies, ranging from publicising polluters and becoming involved in direct action to lobbying for treaty ratification and implementation. In their advocacy and activist roles, NGOs mobilise public opinion in each country as well as globally, often raising alternative visions. The origin of the Global Ozone Observing System, for instance, can be traced to the International Geophysical Year organised by scientists in 1957.

NGOs are in a unique position to link public interest to global and national environmental policy making and implementation. Large transnational federations of national chapters are influential in forming the global environmental agendas with their ability to conceptualise and frame issues. More activist oriented NGOs are involved in monitoring and investigating the record of treaty implementations. Professional organisations such as the International Union for the Conservation of Nature and Natural Resources (IUCN) utilise a scientific knowledge base for mobilising political support. Along with the World Wildlife Fund, the IUCN also accepted administrative responsibilities for the Convention on International Trade in Endangered Species of Fauna and Flora.

Non-state actors complement international organisations in scientific research and policy formulation. With the support of UNEP and UNESCO, the International Council of Scientific Unions (ICSU), a federation of organisations of natural sciences, created, in 1969, the Scientific Committee on Problems of the Environment that studied the impact of human activities on bio-chemical changes on the planet. NGOs have been involved in setting up air pollution standards in Europe, with their support of the European Union's efforts to reduce ozone emission.

NGOs hold their governments and transnational corporations accountable with information gathering and other monitoring activities. Their role is significant especially since there is no central enforcement authority in an international system. Through their watchdog role, NGOs press states to enforce international agreements. NGO monitoring reports can reveal environmentally harmful activities of which the government may have not been aware. Additionally, NGOs represent the concerns of groups of people who are not well represented in national government policies. They often question the norm of free trade in defence of conservation values. In enhancing ecologism, many NGOs pose consistent challenges to neoliberalism.

A coalition made up of NGOs played a particularly important role in preventing large-scale exploitation of marine resources and internal development in the Antarctic by urging states to respect the treaty signed in 1959. The Antarctic and Southern Ocean Coalition (ASOC) established in 1978 represents a worldwide association of over 200 nongovernmental organisations, including Friends of the Earth, Greenpeace and the International Union for the Conservation of Nature. From the early 1980s, Greenpeace and other environmental groups began to lobby to protect the wilderness from mining and other resource extracting activities that endanger the fragile ecosystem in Antarctica (Wapner, 1996, p. 135). Their ideas were supported in the 1991 Environmental Protocol that prohibits mineral exploration for at least fifty years and establishes a framework for preserving the continent as a world park.

FUTURE DIRECTIONS AND OVERVIEW

In spite of shared concerns about the future status of the global environment, there are broad disagreements over appropriate strategies to address environmental degradation. Environmental management is easier in small, culturally homogenous communities with less economic disparity. Uncertainties about the social dimensions of environmental changes increase with disagreement on values and differences in political systems. Due to the fact that population growth, enhancement of technology, and access to resources are uneven both within and across states, it is not easy to impose universal environmental standards. Given the interdependent nature of issue areas and their global scale, however, a high level of co-ordination is required in policy making.

The dynamics of environmental politics is characterised by the fact that no superior authority exists in the global arena to bind everyone to the rules for the proper use of the environment. The question of national sovereignty is at the core of environmental problems since state actors are the major agents for making and implementing national policies (Haas and Sundgren, 1993, p. 407). Regulatory activities can be set up at a multilateral forum, but state actors have to agree to regulate their domestic activities and accept the rules of international regulations.

Getting engaged in voluntary arrangements derives from political and moral interests in responding to long-term environmental deterioration. Many governments pay little attention to environmental matters that exist outside their own borders unless they have a direct impact on them. Joint responses have been made more difficult owing to a lack of consensus and incompatible interests among states. In an international system still dominated by states, the investigation of a record of compliance with environmental regulations is still considered as an infringement of sovereign rights. The attempts to unilaterally control proportions of the global commons have led to introduction of the methods of determining rights of their access and use (Trolldalen, 1992). The formulation of effective responses has to be based on a series of agreements to co-ordinate behaviour and policies of various actors as well as a more general commitment to changes in practices that have destructive ecological consequences.

Global ecological interdependence necessitates modifications to the notion of sovereignty and exclusive territorial control as a constituting principle of international politics (Elliott, 1998, p. 98). Protection of migratory wildlife and prevention of desertification defy solutions within a conventionally bound political jurisdiction. The networks of legal and political accountability are consistent with the norms of interdependency. The lack of motivation and capacity of sovereign states for dealing with an array of environmental problems on their own invites an increased role of the collective international community in defining and applying the accountability obligations.

In addition to competing sources of legitimacy and authority, technological changes, public perception, and cultural diversity all have an impact on the dynamics of environmental politics. The readjustment of human relations with nature, in general, has different political implications, varying in issue areas. Resolving differences can be based on the agreement on normative criteria for rule making, the promotion of ecological values as well as the availability of new scientific knowledge. Environmental principles have often been undermined by the uncompromising principles of free trade which the World Trade Organisation (WTO) advocates over other values. It is well illustrated by the 1998 ruling of the WTO dispute panels which struck down the US law aimed at reducing the mortality of highly endangered sea turtles from foreign shrimp trawling.

In the assumptions of a rational actor model, social interaction is motivated by the assessment of gains incurring from the relationship. In cost-benefit analysis, the policy choice is determined by quantitative measurements of aggregate gains and losses. As the analysis exclusively focuses on economic efficiency represented by the aggregate sum of benefits and costs, whether compensation has to be paid remains a question of distributive fairness (Freeman III, 2000, p. 193). Its logic does not presuppose that benefits would necessarily accrue or be limited to those who pay for meeting a standard. Although objective criteria can be identified in the selection of options, the dilemma of collective action in the environmental arena is how to balance the immediate costs and uncertain future benefits. If the costs to be borne by actions are not compensated by tangible gains in the near future, the sacrifice to secure the collective good is difficult to justify (Rowlands, 1991).

Regime analysis has served as a dominant model for research on global environmental governance over the last decade (Mason, 1999, p. 215). In neoliberal institutionalism, cooperation in collective action situations stems from the expectations of shared gains with adherence to rules agreed upon by community members. With an emphasis on institutional cooperation, functional networks are constructed in narrow domains in the hope that they will eventually bring about structural transformation of the governance system. The incremental path, however, has demonstrated that 'the evidence for deep structural change seems weak at best' (Conca, 1993, p. 310). Response to environmental changes has become increasingly professionalised and bureaucratised. For instance, actions against pollution derive from narrowly defined causes, consequently losing holistic perspectives. On the other hand, the environment is not suitable for simple control strategies (Ayres, 1998, p. 6).

The management of the global environment is not separated from conflict over distribution of power and wealth. Established political conduct has been challenged in broadening our perspectives beyond the existing order supported by prevailing institutions and the market (Brock, 1991). While various devices to ensure procedural fairness have been proposed in promoting voluntary consent, it is a great challenge to bring about the equity of outcomes within a neoliberal institutional framework. In critical approaches (Cox, 1996), environmental issues are examined in connection to the social and political complex as a whole. There is a limitation on exploring the remedies to deal with the structural causes of ecological change within existing disciplinary orthodoxies (Vogler, 1996, p. 13).

Distributing responsibilities for and burdens of improving the conditions of the global environment can be seen in terms of politics of value allocation as well as negotiation of competing interests. The scale of the issues transcends traditional boundaries and exceeds existing institutional capacities. For instance, NGO intervention is located on the local and global continuum as well as the biophysical and political dimensions (Princen, et al., 1994, p. 221). The agreement on the global framework has to be linked to local political and economic solutions.

To the extent that governments can be pressed by advocacy and activist groups, environmental movements both in institutionalised and non-institutionalised forms are able to play an important role in shaping future global agendas. Yet new rules for decision making and institutional forms of co-ordination have to be created in the political realm. This edited volume derives from a continued search for a common ground in exploring a framework for joint efforts.

In Part I 'Values, Norms and Rules', the authors identify specific rules and standards that influence the employment of policy instruments and the implementation of agreements. Marvin Soroos explains various types of rules adopted to regulate activities in the global commons. Based on the distinction between institutional and environmental effectiveness principles, Gabriela Kütting looks at the impact of social systems on the successful implementation of any agreement.

Part II 'Issues in Negotiations' reviews various factors that contribute to successful environmental negotiations. Ho-Won Jeong examines the major characteristics of international environmental negotiation with a particular focus on the processes of agenda setting and bargaining. Pamela Chasek investigates the role of scientific information in changing negotiation dynamics.

In Part III 'Institutional Context', the contributors examine institutional procedures. Rodger Payne looks at how the World Bank's structure and practices were changed in response to the NGO demands for transparency and accountability of the Global Environmental Facility. Beth DeSombre discusses the different types of trade sanctions used for environmental protection.

Chapters in Part IV 'Knowledge, Practice and Politics' deal with a knowledge base, the role of public action, and the formation of new values. Ho-Won Jeong and Charlotte Bretherton explore the ways in which an ecocentric identity can be developed and politicised. Ronnie Lipschutz provides an analytical framework for globalised networks of environmental knowledge and practice.

Overall, this book illustrates a broad spectrum of environmental policy making and institutional practice geared toward the formation of new values and international arrangements. As globalisation influences the transformation of environmental politics, the accumulation of experience requires a systematic analysis of diverse processes and movements. Diverse goals and strategies are likely to be generated by an expanding knowledge base. The rationale of this edited volume is centred on the need for finding an integrative framework to understand the emerging trends and practices of environmental protection.

REFERENCES

R. Ayres, 'Eco-restructuring: The Transition to an Ecologically Sustainable Development', in R. Ayres (ed.), *Eco-restructuring: Implications for Sustainable Development* (Tokyo: United Nations University Press, 1998) pp. 1-52.

H. Breitmeier, 'International Organizations and the Creation of Environmental Regimes', in O. Young (ed.), *Global Governance: Drawing Insights from the Environmental Experience* (Cambridge: MIT Press, 1997) pp. 87-114.

L. Brock, 'Peace through Parks: the Environment on the Peace Research Agenda', *Journal of Peace Research*, vol. 28, no. 4, November (1991) 407-23.

L. Brown, 'Challenges of the New Century', in L. Brown, et al. (eds), *State of the World 2000* (New York: W.W. Norton & Company, 2000) pp. 3-21.

L.K. Caldwell, *International Environmental Policy : from the Twentieth to the Twenty-first Century,* 3rd edn (Durham : Duke University Press, 1996).

J.B. Callicott, 'Toward a Global Environmental Ethic' in N.J. Brown, and P. Quiblier (eds), *Moral Implications of a Global Consensus* (New York: United Nations Environment Programme, 1994) pp. 9-12.

M. Carley and I. Christie, *Managing Sustainable Development* (Minneapolis: University of Minnesota Press, 1993).

N. Choucri, 'Introduction: Theoretical, Empirical, and Policy Perspectives' in N. Choucri, (ed.), *Global Accord: Environmental Challenges and International Responses* (Cambridge: The MIT Press, 1993) pp. 1-40.

K. Conca, 'Environmental Change and the Deep Structure of World Politics' in R. Lipschutz and K. Conca (eds), *The State and Social Power in Global Environmental Politics* (New York: Columbia University Press, 1993) pp. 306-344.

K. Conca, et al., 'Two Decades of Global Environmental Politics' in K. Conca, et al. (eds), *Green Planet Blues: Environmental Politics from Stockholm to Rio* (Boulder: Westview Press, 1995) pp. 5-16.

R. Cox, *Approaches to World Order* (Cambridge: Cambridge University Press, 1996).

L. Elliott, *The Global Politics of the Environment* (New York: New York University Press, 1998).

D.L. Feldman, 'Iterative Functionalism and Climate Management Organizations' in R.V. Bartlett, et al. (eds), *International Organizations and Environmental Policy* (Westport: Greenwood Press, 1995) pp. 187-208.

J. Foster (ed.), *Valuing Nature?: Ethics, Economics and the Environment* (New York : Routledge, 1997).

M. Freeman III, 'Economics, Incentives, and Environmental Regulation', in N. J. Vig and M. Kraft (eds.), *Environmental Policy: New Directions for the Twenty-First Century* (Washington, D.C.: Congressional Quarterly Press, 2000) pp. 190-209.

H. French, 'Coping with Ecological Globalization', in L. Brown, et al. (eds), *State of the World 2000* (New York: W.W. Norton & Company, 2000) pp. 184-202.

P. Haas and J. Sundgren, 'Evolving International Environmental Law: Changing Practices of National Sovereignty', in N. Choucri (ed.), *Global Accord: Environmental Challenges and International Responses* (Cambridge: The MIT Press, 1993) pp. 401-430.

L. Jurgielewicz, *Global Environmental Change and International Law: Prospects for Progress in the Legal Order* (Lanham, Md.: University Press of America, 1996).

S. Kamieniecki and M. Granzeier, 'Eco-Cultural Security and Indigenous Self-Determination', in K. Litfin (ed.), *The Greening of Sovereignty in World Politics* (Cambridge: The MIT Press, 1993) pp. 257-274.

T.W. Luke, 'Sustainable Development as a Power/Knowledge System' in F. Fischer and M. Black (eds), *Greening Environmental Policy* (New York: St. Martin's Press, 1995) pp. 21-32.

M. Mason, *Environmental Democracy* (New York: St. Martin's Press, 1999).

A.S. Mather and K. Chapman, *Environmental Resources* (New York: Longman Scientific & Technical, 1995).

D.H. Meadows, et al., *The Limits to Growth. A Report for the Club of Rome's Project on the Predicament of Mankind* (New York: Universe Books, 1972).

D.H. Meadows, *The Global Citizen* (Washington, D.C.: Island Press, 1991).

C. Merchant, *Earthcare: Women and the Environment* (New York: Routledge, 1996).

K. Milton, *Environmentalism and Cultural Theory* (London: Routledge, 1996).

V.P. Nanda, *International Environmental Law & Policy* (New York: Transnational Publishers, Inc., 1995).

National Research Council, *Global Environmental Change* (Washington, D.C.: National Academy Press, 1992).

OECD (Organisation for Economic Co-operation and Development), *Economic Globalisation and the Environment* (Paris: OECD Publications, 1997).

R. Paehlker, 'Environmental Values for a Sustainable Society', in F. Fischer and M. Black (eds), *Greening Environmental Policy* (New York: St. Martin's Press, 1995) pp. 129-144.

D. Pearce, et al., *Sustainable Development: Economics and Environment in the Third World* (Aldershot: Edward Elgar, 1990).

V. Plumwood, *Feminism and the Mastery of Nature* (London: Routledge, 1993).

G. Porter and J.W. Brown, *Global Environmental Studies* (Boulder: Westview Press, 1996).

T. Princen, 'NGOs: Creating a Niche in Environmental Diplomacy', in T. Princen and M. Finger (eds), *Environmental NGOs in World Politics* (London: Routledge, 1994) pp. 29-47.

T. Princen, et al., 'Transnational Linkages', in T. Princen and M. Finger (eds), *Environmental NGOs in World Politics* (London: Routledge, 1994) pp. 217-236.

J. Rees, 'Market – the Panacea for Environmental Regulation?' *Geoforum*, vol. 23 (1992) pp. 383-94.

R. Rogers, *Nature and the Crisis of Modernity: a Critique of Contemporary Discourse on Managing the Earth* (Montréal: New York: Black Rose Books, 1994).

I.H. Rowlands, 'Ozone Layer Depletion & Global Warming: New Sources for Environmental Disputes', *Peace and Change*, vol. 16, no. 3, July (1991) pp. 260-84.

D. Smith, *Business and the Environment* (London: Paul Chapman Publishing, 1992).

M. Smith, 'To Speak of Trees: Social Constructivism, Environmental Values, and the Future of Deep Ecology', *Environmental Ethics*, vol. 21, Winter (1999), pp. 359-76.

M. Soroos, 'Global Institutions and the Environment', in N. Vig and R. Axelrod (eds.), *The Global Environment* (Washington, D.C.: Congressional Quarterly Press, 1999) pp. 27-51.

D. Stevis, 'Whose Ecological Justice?', *Strategies*, vol. 13, no. 1, 2000, pp. 63-76.

O.S. Stokke, 'Regime as Governance System', in O.R. Young (ed.), *Global Governance: Drawing Insights from the Environmental Experience* (Cambridge: The MIT Press, 1997).

J.M. Trolldalen, *International Environmental Conflict Resolution: The Role of the United Nations* (Geneva: The United Nations Institute for Training and Research, 1992).

J. Vogler, 'The Environment in International Relations', in J. Vogler and M.F. Imber (eds), *The Environment and International Relations* (London: Routledge, 1996).

P. Wapner, *Environmental Activism and World Civic Politics* (Albany: State University of New York, 1996).

World Commission on Environment and Development (WCED), *Our Common Future* (Oxford: Oxford University Press, 1987).

S. Yearley, *Sociology, Environmentalism, Globalization* (London: Sage, 1996).

O. Young, *International Governance: Protecting the Environment in a Stateless Society* (Ithaca, NY: Cornell University Press, 1994).

O. Young, 'Rights, Rules and Resources in World Affairs', in O. Young (ed.), *Global Governance: Drawing Insights from the Environmental Experience* (Cambridge: The MIT Press, 1997) pp. 1-24.

O. Young, *The Effectiveness of International Environmental Regimes* (Cambridge: The MIT Press, 1999).

Part II: Norms and Practice

2 The Evolution of Global Commons

Marvin S. Soroos

The concept of 'commons' is often associated with Garrett Hardin's classic essay 'The Tragedy of the Commons,' although others such as William Forster Lloyd (1933) had written earlier about commons and the tendency for unregulated ones to be overused or misused, even to their destruction (Hardin, 1968). Numerous historical and contemporary examples of commons of a local scale can be identified throughout the world, such as fields or pastures, hunting or fishing grounds, forest tracts, irrigation systems, and aquifers. Studies of many of these commons reveal that it is not unusual for local communities to develop ways of limiting use of these common resources to sustainable levels, thereby avoiding a Hardinian 'tragedy.' In some cases, these local management schemes have operated successfully over many generations (for examples, see Panel on Common Property Resource Management, 1986; Bromley, 1992).

While Hardin uses a parable of the overgrazing of a shared pasture in a mythical English village to explain his theory about the tendency of unregulated commons to be degraded and depleted, he cites examples of commons existing at other societal levels, including some of international and even global scope. At the latter level, several realms are often referred to as commons, namely the oceans (including fisheries), the seabed beyond continental shelves, outer space (including the geosynchronous orbit, the moon and other celestial bodies), Antarctica and surrounding waters, the atmosphere, and the electromagnetic spectrum (Soroos, 1995). These global commons share some basic characteristics with local commons. However, the users of the global commons generally lack the sense of community that frequently exists at the local level, which has complicated the task of developing management systems.

Use of these global commons is to some extent managed by international regimes. The concept 'international regime' has been used widely by scholars and practitioners to describe rearrangements that provide a measure of international governance regarding a specific region, problem, or issue area. Thus, a regime may be comprised of a combination of international institutions, negotiated treaties, customary principles of law, nonbinding resolutions, implementation and enforcement mechanisms, and assistance programs (see Krasner, 1983; Young 1994). There is no overarching international regime that applies to global commons in a general way, but rather an array of regimes that govern specific global commons, such as Antarctica, the oceans, and outer space. Nested within these regimes are subregimes that are even more narrowly focused, such as ones that exist for protection of the stratospheric ozone layer, seabed mining, pollution of the Mediterranean Sea, activities on the Moon, and the conservation of the living resources of the Antarctic region.

These international regimes govern global commons to varying degrees and in different ways, as will be explained in this chapter. Some regimes are much further developed and effective in managing uses of the commons they address than are others. Regardless of their current state of development, most international environmental regimes have proven to be cumbersome and time-consuming to create and strengthen because the participation of states in international institutions is voluntary and, under most circumstances, states cannot be compelled to accept international rules they consider contrary to their interests. Thus, international regimes typically display the lowest common denominator of agreement among nations that have conflicting interests and expectations. The temptation often exists for states to play the role of 'free rider' by refusing to be bound by international regulations pertaining to global commons, while enjoying the common benefits resulting from the willingness of other countries to comply with them.

This chapter provides an overview and analysis of the subject of global commons. The section that follows offers a definition of commons and suggests how it applies to the principal global commons. The essay then proceeds to the question of who owns or has jurisdiction over commons, which may be ambiguous, especially at the international level, but has significant implications for how commons are managed. Next, it compares the international regimes that have been created to manage use of the global commons and the various types of rules that have been adopted to regulate activities in them. The concluding section asks whether there are evolutionary trends in the recognition of global commons and the forms of international governance that apply to them.

DEFINING GLOBAL COMMONS

The concept 'commons' has been defined in a variety of ways that has led to considerable confusion about what qualifies as a commons. The definition proposed here, which is grounded in Hardin's illustration of the pasture in the English village, has three distinguishing characteristics. First, a commons is a domain or collectivity of resources. Second, the entire domain is available to multiple actors who may utilise its resource units for their individual gain. Third, the resource units are both finite and subtractive (see Soroos, 1997, pp. 208-235). Let us now consider each of these attributes in greater detail and reflect on whether they apply to the principal global domains that are often referred to as commons.

The first attribute of commons distinguishes between an encompassing *resource domain*, which is the commons, and the *resource units* that are found within it (Ostrom, 1990, p. 30). In Hardin's analogy, the resource domain is the entire pasture, whereas the clumps of grass that can be eaten by cattle are resource units. Similarly, a forest could be categorised as a resource domain, and individual trees as resource units; a fish stock is a domain while individual fish are the resource units. Other examples of domains and resource units are the deep seabed and polymetalic nodules lying on it; the continent of Antarctica and exploitable resources located within it; outer space, or more specifically the geostationary orbit, and locations where orbiting satellites can be positioned; the atmosphere and specific quantities of air; and the electromagnetic spectrum and specific frequencies used to transmit signals.

These resource domains are of several distinctive types. Antarctica and outer space are spatial domains, whereas the atmosphere and fish stocks are collectivities of physical substances. The oceans could be viewed either as the space occupied by the seas or as the composite of the waters of the oceans, which like air, are constantly moving. The electromagnetic spectrum, sometimes referred to as the 'invisible commons', is a curious domain that is neither a space nor a substance, but a range of frequencies that can be used to transmit electronic signals, or what are known as radio waves, through space without wires.

The second attribute of a commons, that the entire resource domain is available to multiple users for their individual gain, implies that no exclusive claims to any part of a commons are generally recognised. Nor is a resource domain considered a commons. The villagers had access to the entire pasture to graze their privately owned cattle for their personal benefit. A commons may be either *limited access*, which implies there are restrictions or controls on who is permitted to use the commons, or *open access*, which suggests the resource domain is open to any party that desires to make use of it. A commons may be treated as an open-access domain because it is either physically impossible or economically too expensive to deny or restrict access to unauthorised parties. Such domains have been referred to as 'common pool resources' (Ostrom, 1990, p. 30).

Prior to the creation of modern international regimes, the so-called global commons generally were open-access in the sense of being available to all would-be users for their private gain. For centuries under the customary law of the sea, the oceans beyond a narrow coastal band of three nautical miles, or what was known as the high seas, were open to all for navigation and fishing under the doctrine of the 'freedom of the seas.' Likewise, outer space was open to all countries possessing the technical means to launch satellites into orbit; the atmosphere to all who would use it as a medium for disposing of gaseous and particulate pollutants; the airwaves to those who wished to transmit radio signals; and Antarctica to all who desired to conduct research there. In each of these cases, it was the individual users rather than the international community that derived the benefit.

The third defining criterion of a commons is that its resources are both *finite*, meaning that there are limited amounts of them, and *subtractive*, which suggests that resource units exploited by one actor are not available to others. Finite resources are subject to being depleted, either permanently or temporarily depending upon whether they are renewable. By definition, subtractive resources lack the quality of 'jointness,' a concept that means that multiple actors can make use of them without diminishing their value to others (Oakerson, 1992, pp. 43-44). The pasture in Hardin's mythical English village is finite in the sense that the amount of grass that is available for grazing cattle is limited and thus susceptible to depletion until it has had time to grow back. The resources are subtractive in that a clump of grass that is consumed by one cow is no longer available to others.

Fish stocks are clearly both finite and subtractive because fish, though they may be plentiful, are limited in number; and a fish caught and removed from the sea by one boat is no longer available to others. Likewise, while there are immense quantities of mineral-rich nodules lying on the ocean floor, there is a limit to their amount and, after being harvested by one mining company, any given nodule is not available to others. Specific frequency bands on the electromagnetic spectrum will allow only so many clear transmissions of radio signals at any time within a geographical region, depending on the technologies that are used. Each frequency is subtractive in the sense that once it is being used to transmit signals by one party, its simultaneous use by others within the geographical range of the signal of the first will create static, rendering the frequency useless to both. The geostationary orbit is limited and subtractive for much the same reason; if satellites are packed too closely together in the arc, the signals transmitted to and from the satellites will interfere with one another.

The conditions of finiteness and subjectivity are more ambiguous in the case of the atmosphere. Oxygen from the air is removed by terrestrial species, including human beings, through the processes of respiration. While the amount of oxygen in the atmosphere is ultimately finite and specific units consumed by one animal are not available to others, the overall quantities of oxygen are so vast relative to what is removed as to seem infinite. However, the atmosphere is also used as a sink for pollutants, which entails putting substances into it rather than removing them. While the capacity of the atmosphere to take on pollutants, for example, of sulphur dioxide and carbon dioxide, would seem to be without bounds, there are limits to the amount of these pollutants that can be loaded into the atmosphere without causing undesirable environmental changes, such as acidification and global warming. Moreover, once some polluters have helped themselves to portions of the atmosphere's capacity to absorb pollutants without causing these changes, they are no longer available to would-be polluters. Thus, as a sink for pollutants without harmful consequences, it could be argued that the atmosphere is both finite and subtractive. The ozone layer and climate, which could also be looked upon as atmospheric resources, are neither finite nor subtractive in that there is no limit to the number of people who can derive benefits from them, and one party's enjoyment of them does not detract from what is available to others. Such resources are known as 'public goods' (Olson, 1965).

OWNERSHIP AND JURISDICTION

The preceding section defines a commons in terms of the attributes of the resources that it encompasses and their accessibility to individual users for their personal gain. It makes no assumptions about ownership of the domain, although in most cases who owns the domain has implications for who has jurisdiction over it, and accordingly the prerogative of making the rules about its use. In the case of Hardin's village, it is often assumed that the pasture is owned by the community as a whole, in which case it would be considered *common property*. There are other possibilities, however. The pasture could be the property of an absentee landowner who is benevolent or disinterested enough to permit it to be used by the villagers. Likewise, it could be the land of the king or of the state, who permits the local villagers to use it as a commons. Finally, there may be no recognised owner of the land.

Most local commons have recognised owners, but for most global commons, the situation is either undefined or ambiguous. Usually, it is a question of whether the resource domain is unowned or, alternatively, belongs to the entire international community. Unowned domains can be commons if it is generally understood that they cannot be claimed in whole or in part by any individual actors, as was commonly done by explorers who discovered new lands centuries ago. Even if a domain is unowned, the users can band together to adopt a regulatory scheme. They may decide to exclude others, thus making the commons a limited-access domain, or to impose restrictions on their own use of it. Domains considered to belong to the international community as a whole are sometimes declared to be the 'common heritage of mankind.' The common heritage designation implies that a domain is to be used exclusively for peaceful purposes and that all states have a right to share in the benefits derived from the domain, regardless of whether they currently possess the technological and economic means to exploit its resources. Furthermore, it is presumed that all states, as partial owners of the domain, are entitled to participate in the decision-making processes that establish regulations about its use (Birnie and Boyle, 1992, pp. 120-21).

The high seas are perhaps the purist example of a resource domain that is unowned and not available for national claims. This legal status of the high seas has not always been respected, as when Portugal and Spain concluded the Treaty of Tordesillas in 1494, which drew an imaginary pole-to-pole line of demarcation in the Atlantic Ocean 370 leagues west of the Cape Verde Islands. Portugal was given exclusive rights to unclaimed lands and ocean commerce to the east of the line, which included the Indian Ocean, and Spain to such areas to the west of the line, including the Pacific Ocean (Fulton, 1911, pp. 106-7). These exorbitant claims were vigorously challenged by the other maritime powers of that era, including England and the Netherlands, and later contested by Dutch legal expert Hugo Grotius in his seminal treatise *Mare Liberum* (1609), which reasserted the 'freedom of the seas' doctrine that has prevailed to this day.

In the decades following the Second World War, as many as 100 coastal states infringed on the high seas by unilaterally extending their claims to ocean areas off their coasts beyond the traditional three-mile limit to as far out as 200 miles, in order to exclude other states from nearby fisheries or oil and gas resources in the continental shelf (Barkenbus, 1979, p. 29). In many cases, these expanded claims were precipitated by the threat that huge distant-water fishing fleets, built up by countries such as the Soviet Union, Poland, Japan, and Spain posed to the coastal fisheries upon which local communities had traditionally depended for their livelihoods. These claims were eventually sanctioned by the Convention on the Law of the Sea of 1982, which gave coastal states the primary right to the resources in exclusive economic zones (EEZs) that extended out 200 nautical miles from their shorelines. While coastal states were not given the full prerogatives of sovereignty over the EEZs, they were in effect granted control over the harvesting of fish stocks that are concentrated in these areas. Complications arise, however, in determining the legal status of shared stocks that move among the EEZs of several states or straddling stocks that move out from EEZs into the high seas. Under terms of the Law of the Sea Convention, states harvesting such stocks are expected to cooperate in establishing rules to prevent overfishing. Most states (Norway and Japan being notable exceptions) have accepted the regulations of the International Whaling Commission on the harvesting of whales even though numerous whale species reside at least part of the time in the coastal waters of states (see Stoett, 1997; Skåre 1994 on the rationale for Norwegian whaling).

The seabed, which includes the ocean floor and subsoil beyond continental shelves, is the leading example of a domain that is subject to the 'common heritage' principle. The terminology was first proposed in 1967 by Arvid Pardo, the United Nations delegate from Malta, who was alarmed about claims being made by some states to large sectors of the deep seabed on the basis of their geographical proximity. Advocated by Third World countries lacking the economic and technological capacity to exploit the resources of the deep seas, the common heritage language was incorporated into a declaration of principles on the seabed adopted by the UN General Assembly in 1970 and later in the text of the 1982 Law (Hollick, 1981).

The Law of the Sea Convention explicitly declares that rights to the resources of the seabed, or what is known as the Area, 'are vested in Mankind as a whole.' Moreover, no state is permitted to extract the resources of the Area except in accordance with the conditions set forth by an International Seabed Authority (ISA), which was to be established to carry out the provisions of the convention. The economic benefits of activities undertaken in the Area would be equitably shared by all peoples taking into account the specific needs of developing countries and non-self-governing peoples. To achieve this objective, the Law of the Sea Convention provided for the creation of an International Enterprise under the auspices of the ISA, which would mine the resources of the seabed. In the negotiations on the convention, Third World countries initially sought to limit all seabed mining to the Enterprise, but compromised on a 'parallel' system that also allowed nation-based mining companies to exclusive rights to mine certain tracts of the ocean floor (see Churchill and Lowe, 1983). It was these provisions pertaining to the seabed that prompted the United States and several other industrial countries to refuse to sign or ratify the Law of the Sea Convention.

The legal status of outer space is more ambiguous. The Outer Space Treaty of 1967 does not address the question of ownership of outer space, but is unequivocal in declaring that no part of outer space, including the moon and other celestial bodies, may be appropriated by any nation. The only notable challenge to this provision was mounted in 1976 when eight equatorial states adopted the Bogotá Declaration, in which they asserted exclusive claims to the sections of the geostationary orbit that lie over their territories. In making this claim, the equatorial countries contended that the geostationary orbit was a natural extension of their national air space, the upper boundary of which had not been defined in international law, and that satellites revolving around the Earth in this orbit remained above the same country. Most other countries have refused to recognise the claims of the Bogotá group to parts of the geostationary orbit (Gorove, 1979, pp. 450-55).

The Outer Space Treaty, which was adopted the same year as Pardo's United Nations speech on the seabed, has provisions that are to some extent suggestive of the common heritage principle.

The treaty's first article declares that the 'exploration and use of outer space, including the moon and other celestial bodies, shall be carried out for the benefit and in the interests of all countries, irrespective of their degree of economic or scientific development, and shall be the province of all mankind'. The article goes on to suggest, however, that outer space and celestial bodies 'shall be free for exploration and use by all states without discrimination of any kind,' but in the next clause provides that these activities shall be conducted 'on a basis of equality'. Third World countries have argued that these provisions oblige developed countries to take steps that make it possible for them to become partners in the exploration and use of outer space. The Moon Treaty of 1979 goes further in explicitly declaring that the moon and other celestial bodies are the 'common heritage of mankind'. Accordingly, resources from them cannot be appropriated by any state for its exclusive use, but must be conserved or exploited for the benefit of all (Birnie and Boyle, 1992, p. 120; see also Christol, 1982).

Antarctica's legal status has also been ambiguous, but for different reasons. Seven countries – Argentina, Australia, Chile, France, New Zealand, Norway, and the United Kingdom – have long-standing claims to pie-shaped segments of the continent extending out from the South Pole, some of which are overlapping, while a small part of the continent remains unclaimed. No other countries, including the United States and the Soviet Union have recognised these national claims to Antarctica. The status of the previous claims was put on hold by the Antarctic Treaty of 1959, which neither recognised or nor denied them. The treaty did, however, rule out any new or expanded claims. The twelve countries with significant research activities on the continent that negotiated the treaty and became its initial parties formed a group known as the Antarctic Treaty Consultative Parties (ATCP). The ATCP has grown to 26 states with the induction of new members that have developed a significant research presence on the continent. The group initially met biannually, but in recent years annually, to consider additional agreements pertaining to the region and its environment (see Peterson, 1988).

The Antarctic treaty system, and in particular the role played by the exclusive ATCP in making international policies for the

Antarctic region, was challenged in the United Nations during the 1980s. Concerned about the prospect that the ATCP would adopt rules permitting exploration and exploitation of the minerals of Antarctica, a group of Third World countries led by Malaysia and Antigua and Barbuda sought to have the Antarctic region declared the 'common heritage of mankind'. Thus, as with the seabed, under this proposal the principle would be established that all countries were entitled to share in whatever wealth could be generated from the Antarctic region and participate in decisions on the region's management. UN General Assembly resolutions introduced to effect such a change failed in the face of opposition from the ATCP, whose Third World membership had grown with the addition of Brazil, China, India, Peru, and Uruguay to the ATCP charter members Argentina and Chile (Peterson, 1988, 175-93).

The relatively small number of treaties that apply to the atmosphere offers little guidance on its legal status; in fact, the term 'atmosphere' is rarely mentioned in any international law agreements. One consideration is that approximately 85 percent of the air comprising the atmosphere lies in its lowest layer, the troposphere, which extends upward from the Earth's surface to an altitude varying from 8 to 20 km. The air circulating within the troposphere moves through what is known as *air space*, which is a three-dimensional area that lies between the surface of the earth and outer space.[1] The applicable international law on air space – the 1944 Chicago Convention on International Civil Aviation – provides that a state has 'complete and exclusive sovereignty over the airspace above its territory', which includes its land area and adjacent territorial waters (Abeyratne, 1996, pp. 1-11). Thus, states would seem to have exclusive rights to the air residing within their air space at any given time, while air that is located over areas that are not part of the any state's territory would presumably be unowned and thus have a status comparable to the high seas. Thus, any given unit of air could be said to be owned for the time when it is in the air spaces of countries, and unowned at other times when it is above non-national areas.

States have not been possessive of the air above them, apparently because it is bountiful and perpetually moving so that the specific quantities of air above them are always changing.

Furthermore, it would be impossible for states to enclose substantial amounts of air for their exclusive use. However, they have not been willing to relinquish all claims to the atmosphere and declare it a global resource. Being a small island state threatened by rising seas from projected global warming, Malta proposed in the General Assembly in 1988 that the global climate be designed the 'common heritage of mankind', as it had done for the seabed twenty years later. Malta's suggestion was not adopted, but in its place, the UN General Assembly adopted a resolution which refers to climate change as a 'common concern of humankind.' This terminology which appears again in the first clause of the 1992 Framework Convention on Climate Change (see Bodansky, 1993, p. 465). It has also been applied to the conservation of biological diversity in another convention adopted on that subject at the Earth Summit. The concept is a rather new one in international law and as yet is not well defined. At a minimum, it can be assumed that subjects so declared have a place on the agendas of international bodies and can be viewed as a legitimate subject of international attention (Boyle, 1991, p. 11). When applied to resources within national jurisdictions, 'common concern' does not appear to imply that the international interest infringes upon sovereign prerogatives.

Ownership of the electromagnetic spectrum has also not been explicitly clarified, although the International Telecommunications Union has referred to it as a common heritage resource (Savage, 1989, p. 20-21). It has generally been presumed that states have the sovereign prerogative of regulating uses of the spectrum for transmitting signals within their borders. Because of the threat of interference and the possibility of chaotic static on the airwaves, states have been willing to accept rules established by the International Telecommunications Union that limit their use of the air waves for transmissions that go beyond their borders. For the most part, they have refrained from staking permanent claims to the specific frequencies they use internationally, but in practice, their continued control of them is rarely challenged.

REGIMES FOR GLOBAL COMMONS

One or more international regimes have been created to govern use of each of the major global commons. These regimes have evolved in one of two ways. One path has been the creation of a *general framework* regime that provides a structure for the management of the entire commons, followed by the adoption of more narrowly focused, problem-oriented sub-regimes. The other path commences with the establishment of a *problem-oriented* regime which only later may lead to the development of a comprehensive regime for the commons. The regimes created for Antarctica and outer space followed the first path; the ones that govern the oceans, the atmosphere, and the electromagnetic spectrum the second path.

The framework agreement for the Antarctic regime is the Antarctic Treaty of 1959, which was negotiated in the aftermath of the International Geophysical Year of 1957-58 primarily to circumvent potential military conflict in the region and to ensure that the entire area would remain open to scientific research. The Antarctic Treaty Consultative Parties (ATCP) proceeded to establish more specific rules designed to conserve the species and environment of the region, including the Agreed Measures for Antarctic Flora and Fauna (1964), the Convention for the Conservation of Antarctic Fur Seals (1972), and the Convention on the Conservation of Antarctic Marine Living Resources (1980). The ATCP then took up the subject of exploration and mining of minerals in Antarctica, agreeing first to the Convention on the Regulation of Antarctic Mineral Resource Activities (CRAMRA) in 1989, which would have permitted mining activities, but with substantial environmental safeguards. However, blocked from coming into force by the reservations of Australia and France, CRAMRA was pre-empted two years later by the Madrid Protocol on Environment Protection to the Antarctic Treaty, which is designed to minimise the human impact on the region and in effect to preserve it as a world park. It imposes a ban on commercial mining for minerals and drilling for oil and natural gas in the region, which for the next fifty years can be altered only by a unanimous vote of the ATCP (Stokke, 1992).

The Outer Space Treaty of 1967, which similarly was the first international treaty pertaining to the use of outer space, provided the foundation for the development of a regime for the region. It was negotiated in the UN Committee on the Peaceful Uses of Outer Space (COPUOS), which had been established soon after the Soviet Union launched its first Sputnik satellite into earth orbit in 1957. As with the Antarctic Treaty adopted eight years earlier, the Outer Space Treaty limited military activities, including a ban on the positioning of nuclear weapons or other weapons of mass destruction in outer space, while the moon and other celestial bodies are to be used exclusively for peaceful purposes. The Outer Space Treaty also resembles the Antarctic Treaty in affirming the freedom of scientific activities and encourages international scientific cooperation. Several more narrowly focused agreements followed, including ones on the rescue and return of astronauts and space objects (1969), international liability for damage caused by space objects (1972), and the registration of satellites and other orbiting anthropogenic objects (1976). The more restrictive, and thus controversial Agreement Concerning the Activities of States on the Moon and other Celestial Bodies was adopted in 1979, but has not yet been ratified by the major space powers (see Christol, 1982).

In the case of the oceans, numerous problem-oriented regimes were created prior to the adoption of the comprehensive Law of the Sea Convention in 1982. Among these were thirteen international fishery commissions, four of which were designed to conserve specific species of fish or marine mammals, namely fur seals, whales, tuna, and salmon, while the remaining nine dealt with managing stocks of multiple species of fish within certain geographical regions of the oceans, examples being the northeast Atlantic, the northern Pacific Atlantic, and the Baltic Sea (see Peterson, 1993). Pollution problems have triggered the formation of a number of regimes, such as the extensive rules of the International Maritime Organisation (formerly the Intergovernmental Maritime Consultative Organisation) on vessel-source oil pollution, the first having been adopted in 1954 (see Mitchell, 1994). The Oslo and London Conventions of 1972 respectively regulate the dumping of toxic substances in the North Sea and the oceans generally. Regimes have also been adopted to limit pollution concentrated in semi-enclosed seas, the first of which were for the Baltic Sea (1974) and Mediterranean Sea (1976).

The Law of the Sea Convention was the culmination of an effort begun by the United Nations International Law Commission (ILC) in 1950 as a growing number of coastal states were unilaterally claiming broader zones off their shores than were recognised under the customary three-mile limit that had prevailed for centuries. The ILC drew up separate draft treaties on territorial seas and contiguous zones, continental shelves, the high seas, and fishing and the conservation of living resources, which were adopted at the first UN Conference on the Law of the Sea (UNCLOS I) in 1958. UNCLOS I failed, however, to reach an agreement on perhaps the most important issue before it, the breadth of the territorial seas that could be claimed by coastal states, as did a second Law of the Sea Conference (UNCLOS II) held in 1960 (Juda, 1996, pp. 138-59).

Concern over Pardo's warnings about exorbitant national claims to the deep seabed prompted the United Nations to convene a third Law of the Sea Conference (UNCLOS III), with sessions being held intermittently between 1973 and 1982. As initially proposed in 1967, UNCLOS III would have focused primarily on drawing up a regime for the seabed, but its agenda was soon expanded to drafting a comprehensive law of the sea. The resulting convention finally resolved the issue of the breadth of coastal states' control over waters and continental shelf adjacent to their coasts. Territorial waters would now extend out to 12 nautical miles and an exclusive economic zone (EEZ) to 200 miles. As a result, numerous fish stocks that had once been pure examples of international commons now came under the jurisdiction of coastal states, which in most cases treat them as national commons that too soon become badly depleted.

The seabed beyond continental shelves did not become a salient international concern until the 1960s, when it first appeared that technologies under development might make it possible to profitably mine the mineral-rich seabed nodules. There were still no problem-focused treaties pertaining to the use of the seabed by the time of the adoption of the Law of the Sea Convention. The controversial Part XI of the lengthy Law of the Sea Convention laid out the framework for a seabed regime, which is separate and distinct from the regime that governs use of the high seas. Having been declared a common heritage of mankind, the seabed is subject to more elaborate and restrictive rules on the exploitation of its resources, which are to be implemented by the new International Seabed Authority. However, it was not until 1994 that the convention received the ratifications of sixty states required for it to come into force. This was accomplished only after a supplemental agreement was reached that the industrial countries considered less prejudicial to their interests in mining the seabed (see Leitner, 1996). Thus, with its origins in a framework agreement, the evolution of the seabed regime parallels the development of the Antarctic and outer space regimes, which also govern remote domains where there had been relatively little prior activity (Juda, 1996, pp. 209-54).

The atmosphere has not been the subject of an overarching regime, but several problem-oriented regimes have been created independently of one another. The first of the atmospheric regimes was based on a single international agreement, the Partial Test-Ban Treaty of 1963, which banned testing of the nuclear weapon in the atmosphere, oceans, and outer space. Three other regimes have been developed using the framework/protocol approach. The problem of acid deposition in the European region is addressed by the Long Range Transboundary Air Pollution (LRTAP) regime, which includes a framework treaty (1979) and protocols on sulphur (1985, 1994), nitrogen oxides (1988), and volatile organic compounds (1991). The threat to the ozone layer posed by chlorofluorocarbons (CFCs) and other human-produced chemicals is thoroughly dealt with by a remarkable regime based on the framework Vienna Convention (1985), a rather weak document that was supplemented by the Montreal Protocol (1987) and major revisions that were adopted in London (1990), Copenhagen (1992), and Vienna (1995). Finally, the Framework Convention on Climate Change, concluded at the Earth Summit in 1992 has been supplemented by the Kyoto Protocol (1997), which commits the developed countries to significant net reductions of greenhouse gas emissions. Despite these efforts to address atmospheric problems related to polluting activities, there has been virtually no movement toward negotiating a comprehensive law of the atmosphere (see Soroos, 1997).

Specialised regimes have been developed to manage use of the electromagnetic spectrum and the geostationary orbit. International assignments of certain portions of the electromagnetic spectrum, known as bands, were first made to various types of communication activities at a radio conference held in Berlin in 1906. The first comprehensive assignment plan for the spectrum was adopted in 1927, which underwent a major revision in 1947 soon after the International Telecommunications Union was recognised as a specialised agency of the United Nations. At that time, the International Frequency Registration Board was created to implement the plan and maintain a registry of national uses of specific frequencies consistent with the rules of the management scheme, which offers authorised registered users some protection against interference from other users. Rules pertaining to the use of the spectrum were renegotiated at general World Administrative Radio Conferences (WARCs) of the International Telecommunications Union held in 1959 and 1979. With the rise in use of the geostationary orbit, plans were adopted that allotted frequencies for use in sending information to and from communication satellites for the European, Asian, and African regions in 1977 and for the western hemisphere plans for spaces in the geostationary orbit were adopted at specialised WARCs held in 1985 and 1988 (see Zacher, 1996, pp. 127-80).

STRATEGIES FOR MANAGING THE USE OF GLOBAL COMMONS

The international regimes that govern global commons were established for a variety of reasons, among which was the need to regulate human activities that were already having, or could have, undesirable environmental impacts. These problems are of a variety of types, including depletion of resource units (such as collapse of marine fisheries), contamination (such as acidification caused by transboundary air pollution), environmental disruptions (such as depletion of the ozone layer and climate change), and interference and overcrowding (such as static on the airwaves and crowding of the geostationary orbit). Numerous types of regulations have been adopted by the international regimes that govern these domains, most of which take the form of prohibitions, limits, rules on equipment or operating procedures, or the assignment of exclusive rights.

Prohibitions are normally enacted when the overriding objective is to preserve a resource domain in its uncontaminated condition, or when depletion of its resource units has reached a point that drastic steps are needed to enable them to recover. The International Whaling Commission enacted a moratorium on the commercial harvesting of whales throughout the oceans, which took effect in 1986 at a time when stocks of most whale species had declined to small fractions of their virgin numbers and some were threatened by extinction. The Partial Test-Ban Treaty of 1963 banned the testing of nuclear weapons in three commons–the atmosphere, outer space, and oceans. The Oslo and London Conventions of 1992 proscribed the use of the oceans as a disposal site for certain toxic substances, such as mercury, cadmium, DDT, PCBs, high-level radioactive wastes, and agents of chemical and biological warfare, which appear on a 'black list' that is periodically updated. The 1990 London and 1992 Copenhagen amendments to the 1987 Montreal Protocol set dates after which CFCs, halons, and several other substances that threaten the ozone layer may not be produced or traded. Finally, the 1991 Madrid Protocol to the Antarctic Treaty indefinitely rules out all mining activities on the frozen Southern continent in order to maintain its pristine and fragile environmental qualities. When such prohibitions are enacted, the domain no longer fulfils the definition of a commons, at least for the banned uses.

Other management schemes are designed simply to limit the extent to which a commons is used, usually to prevent or lessen the amount of environmental damage that takes place. It was once a common practice of international fishery commissions to impose a universal quota, which upon being filled by the combined catch of the boats from all states, would trigger a closing of the fishing season. The tendency for universal quotas to stimulate an over-investment in fishing boats led many of the commissions to allocate a certain amount of the total allowable catch to each of the member nations. The grounds for making the allocations have been contentious issues, which usually involve factors such as the geographical proximity of countries and their historical shares of the catch. The most common approach for limiting air pollutants has been to mandate that states freeze their emissions at the level of a recent year, as was done in the 1988 Nitrogen Protocol of the LRTAP regime; or reduce them by an across-the-board percentage, an early example being the LRTAP regime's Sulphur Protocol of 1985 which required the parties to reduce sulphur dioxide emissions by 30% (from 1980 levels) by 1993. Similarly, the original Montreal Protocol of 1987 required the parties to reduce CFC production by 20% by 1993 and by 50% by 1998, with developing countries being given a 10-year grace period and technical and economic assistance in adopting substitutes for the controlled substances. The Revised Sulphur Protocol of 1994 is a more sophisticated regulatory instrument in that each state was given an individualised goal for emission reductions based on its emissions of pollutants and the countries where they are deposited, taking into account the recipient country's 'critical load' for acidic deposition without serious environmental effects.

Some of the regimes have established rules about what equipment and operating procedures may or may not be used in order to lessen environmental damage or interference. Fishery commissions have specified minimum mesh sizes for nets to allow smaller specimens to escape and grow to maturity. In recent years, efforts have been made in the United Nations to prohibit the use of drift nets (some of which are as long as 50 kilometres) that are having a devastating impact on marine life. The International Maritime Organization has mandated that oil tankers be constructed with segregated ballast tanks and use 'load on the top' procedures to reduce intentional discharges of oil substances. Certain types of navigational equipment are required as a measure designed to avoid collisions and groundings that can cause massive oil slicks. The International Telecommunications Union has required the use of equipment, such as single side-band transmitters for sending high frequency (HF) signals, which has reduced interference and allowed for more efficient use of both the electromagnetic spectrum and for the geostationary orbit.

In situations in which multiple users of a domain may cause interference with one other, the practice has sometimes been to give users exclusive rights to use designated parts of it. Such has been the case with both frequencies in the electromagnetic spectrum and spaces in the geostationary orbit. At first, the initial presumption was that the first state to use specific frequencies within a geographical area or positions in the geostationary orbit had a right to continue using them without interference from others, although this was not interpreted as a permanent allocation. This 'first-come, first served,' or *a posteriori*, arrangement for establishing the continuing right to use frequencies and orbital slots has been challenged by Third World countries out of concern that by the time they were ready to use these resources domains, they would already be fully occupied by the more technologically advanced countries. Thus, the developing countries pushed for *a priori* allocations of parts of the radio spectrum and geostationary orbit, which would reserve spaces for each country in these domains regardless of whether they were ready to use them (Soroos, 1982). Such allocations have been made for some frequency bands, including those to be used for satellite communications, and the geostationary orbit. The seabed regime provides that state-based mining companies can be granted exclusive rights to mine certain tracts of ocean floor, but only on the condition that they provide several types of assistance for an international mining company, known as the Enterprise. State-based companies are not only required to identify a comparable seabed tract to be reserved for the Enterprise, but also to share their seabed mining technologies at a reasonable price.

TRENDS IN GLOBAL COMMONS

At the international and global levels, the idea of commons is more of an analytical concept than a legal one. There is no international law of commons that applies generally to the global commons that have been discussed in this chapter. The concept is useful as an analytical tool for identifying a number of resource domains that share certain attributes and for appraising the strategies that have been used to regulate human uses of them. The preceding overview and analysis suggests several trends in global commons and how they are managed.

1. The number of global commons has increased as humanity has developed the capacity to explore and make use of remote realms such as Antarctica, outer space, and the deep seabed.

2. States have unilaterally staked exclusive claims to portions of several of the global commons, including coastal waters and continental shelves, the deep seabed, and the geostationary orbit, but only in the case of expanded coastal areas have these claims been acknowledged by other states and eventually legitimised by international treaties.

3. The freedoms that states once enjoyed to make use of commons such as the oceans, fisheries, the electromagnetic spectrum, outer space, and the atmosphere have gradually been circumscribed by the regulations of international regimes.

4. A variety of circumstances have led to the development of international regimes to govern activities in international commons including: (a) the assertion of national claims (Antarctica and the seabed), (b) the prospect of development or resource extraction (outer space, the seabed, and Antarctica), (c) manifestations of environmental damage (air pollution), (d) depletion of renewable resources (fisheries), (e) interference (electromagnetic spectrum), and (f) compelling scientific evidence of future environmental change (ozone layer and climate change).

5. International regimes that govern the global commons were for the most part developed independently of one another, although some states have attempted to transfer principles or policies from one regime to others, such as the 'common heritage of mankind' doctrine.

6. The regimes that govern some of the commons originated from framework agreements that laid out general principles and mechanisms for supplementary agreements (Antarctica, outer space, seabed), whereas others proceeded from more focused, problem-oriented regimes (oceans, atmosphere, electromagnetic spectrum).

7. Considerable diversity exists in the rules that have been developed to manage use of the global commons, although there is some consistency across those adopted by similar types of regimes; for example, universal and national quotas (fishery regimes), freezes and across-the-board cut backs (LRTAP and ozone layer regimes), and exclusive user rights (electromagnetic and geostationary orbit regimes).

8. Increasingly, developing countries have been raising the issue of equity in the use of international commons, for example in pushing for the application of the common heritage principle to additional regimes (seabed, moon, and other celestial bodies) and for the adoption of *a priori* as opposed to *a posteriori* allocation schemes (airwaves and geostationary orbit).

9. Regimes that govern activities in spatial domains (oceans, seabed, Antarctica, and outer space) have evolved further than those that are comprised of physical substances which pass through areas under the jurisdiction of states (atmosphere).

10. Regimes that have prohibited activities involving global commons (Antarctic minerals, atmospheric test ban, ozone depletion) have generally been more successful than those that have merely sought to limit them (fisheries, outer space, LRTAP), which require resolving additional issues of fairness and creating more complicated provisions for enforcement.

11. Under the rules of a number of the regimes, resource domains are no longer commons for certain uses by the definition adopted in this article, as when there are prohibitions on using them or when parts of them are allocated to individual states temporarily (seabed, electromagnetic spectrum, geostationary orbit) or permanently (coastal fisheries).

CONCLUSIONS

This chapter has made the case that there are commons at the global level that share the basic attributes of commons existing at other levels of human society: a domain encompassing resource units, available to multiple users for individual gain, and the physical qualities of being finite and subtractive. At the global level the commons system was neither invented nor created, commons simply came into being as human activities expanded natural resource domains beyond the jurisdiction of sovereign states, such as the seas, Antarctica, and outer space. The continued existence of global commons have been challenged by two types of developments: first, exclusive claims staked by states to parts of the commons and, second, various rules incorporated into international regimes that prohibit uses of entire commons, or parts of them. Some relatively pure examples of global commons remain, such as the high seas, most of outer space, and the atmosphere; while other domains including the seabed, Antarctica, the moon, the electromagnetic spectrum, and the geostationary orbit are not as clear-cut examples of commons.

This chapter has called attention to the diversity of global commons and the international regimes that have arisen to manage their use. It has touched only lightly on the adequacy of the international arrangements that have been created for managing uses of the global commons and the complications that have arisen in negotiating these arrangements among the governments of sovereign states. The process of reaching broad agreement on international treaties tends to be notoriously slow, with the resulting documents often reflecting the lowest common denominator of common interests among diverse states. And, as with commons at other levels, there is the ever-present challenge of coping with would-be 'free riders,' who are reluctant to submit to rules on the use of commons, even as they take advantage of the restraint of others. Even among states that formally agree to be bound by rules on the use of commons, compliance is often a problem in view of the absence of effective international enforcement mechanisms. Nevertheless, there are examples of effective international regimes for global commons, such as those that ban nuclear tests in the atmosphere, preserve the stratospheric ozone layer, and regulate use of the airwaves and geostationary orbit. By contrast, other regimes have been notably less effective, such as those that seek to regulate marine fishing, or are still in an early stage of development, an example being the one that addresses the looming threat of global climate change.

NOTES

1. While the boundary between air space and outer space has not been defined, it is generally understood that air space includes the altitudes at which aircraft fly, while outer space is the realm of artificial satellites. Since planes can fly at upwards of 30 km., air space includes all of the troposphere and a major part of the stratosphere (Christol, 1982, pp. 435-546).

REFERENCES

R.I.R. Abeyratne, *Legal and Regulatory Issues in International Aviation* (Irvington-on-Hudson, NY: Transnational Publishers, 1996).

J.N. Barkenbus, *Deep Seabed Resources: Politics and Technology* (New York: Free Press, 1979).

P.W. Birnie, and Alan E. Boyle, *International Law and the Environment* (New York: Oxford University Press, 1992).

D. Bodansky, 'The United Nations Framework Convention on Climate Change: a Commentary', *Yale Journal of International Law*, Vol. 18, No. 2, (1993) 451-558.

A. Boyle, 'International Law and the Protection of the Atmosphere', in Robin Churchill and David Freestone (eds), *International Law and Global Climate Change* (London: Graham & Trotman, 1991) pp. 7-20.

D.W. Bromley, (ed.), *Making the Commons Work: Theory, Practice, and Policy* (San Francisco, CA: ICS Press, 1992).

C. Christol, *The International Law of Outer Space* (New York: Pergamon, 1982).

R.R Churchill, and A.V. Lowe, *The Law of the Sea* (Manchester: Manchester University Press, 1983).

T.W. Fulton, *The Sovereignty of the Sea* (Edinburgh: William Blackwood, 1911).

S. Gorove, 'The Geostationary Orbit: Issues of Law and Policy', *American Journal of International Law* Vol. 73, No. 3, July, (1979) 444-61.

G. Hardin, 'The Tragedy of the Commons', *Science*, Vol. 168, No. 3859, December 13, (1968) 1243-48.

A.L. Hollick, *U.S. Foreign Policy and the Law of the Sea* (Princeton, NJ: Princeton University Press, 1981).

L. Juda, *International Law and Ocean Use Management: The Evolution of Ocean Governance* (New York: Routledge, 1996).

S.D. Krasner, ed., *International Regimes* (Ithaca, NY: Cornell University Press, 1983).

P.M. Leitner, *Reforming the Law of the Sea Treaty: Opportunities Missed, Precedents Set, and U.S. Sovereignty Threatened* (Lanham, MD: University Press of America, 1996).

W.F. Lloyd, *Two Lectures on the Checks to Population*, 1968 reprint (New York: Augustus M. Kelley, 1833).

R.B. Mitchell, *Intentional Oil Pollution at Sea : Environmental Policy and Treaty Compliance* (Cambridge, MA: MIT Press, 1994).

R.J. Oakerson, 'Analyzing the Commons: A Framework', in Daniel W. Bromley, ed., *Making Commons Work: Theory: Practice, and Policy* (San Francisco, CA: Institute for Contemporary Studies, 1992) pp. 41-59.

M. Olson, *The Logic of Collection Action: Public Goods and the Theory of Groups* (New York: Schoken Books, 1965).

E. Ostrom, *Governing the Commons: The Evolution of Institutions for Collective Action* (New York: Cambridge University Press, 1990).

Panel on Common Property Resource Management, *Proceedings of the Conference on Common Property Resource Management*, (Washington, DC: National Academy Press, 1986).

M.J. Peterson, *Managing the Frozen South*, (Berkeley, CA: University of California Press, 1988).

M.J. Peterson, 'International Fisheries Management', in Peter M. Haas, Robert O. Keohane, and Marc A. Levy, (eds), *Institutions for the Earth: Sources of Effective International Environmental Protection* (Cambridge, MA: The MIT Press, 1993) pp. 249-308.

J.G. Savage, *The Politics of International Telecommunications Regulation* (Boulder, CO: Westview Press, 1989).

M. Skåre, 'Whaling: A Sustainable Use of Natural Resources or a Violation of Animal Rights?' *Environment*, Vol. 36, No. 7, September, (1994) 12-20, 30-31.

M.S. Soroos, 'The Commons in the Sky: the Radio Spectrum and Geosynchronous Orbit as Issues in Global Policy', *International Organization*, Vol. 36, No. 3, Summer, (1982) 665-677.

M.S. Soroos, 'The Tragedy of the Commons in Global Perspective', in Charles E. Kegley and Eugene R. Wittkopf, (eds), *The Global Agenda: Issues and Perspectives* (4th edition) (New York: McGraw Hill, 1995) pp. 422-435.

M.S. Soroos, *The Endangered Atmosphere: Preserving a Global Commons* (Columbia, SC: University of South Carolina Press, 1997).

P.J. Stoett, *The International Politics of Whaling* (Vancouver: UBC Press, 1997).

O.S. Stokke, 'Protecting the Frozen South', in *Green Globe Yearbook 1992* (New York: Oxford University Press, 1992) pp. 133-40.

O.R. Young, *International Governance: Protecting the Environment in a Stateless Society* (Ithaca, NY: Cornell University Press, 1994).

M.W. Zacher, *Governing Global Networks: International Regimes for Transportation and Communication* (Cambridge: Cambridge University Press, 1996).

3 New Dimensions of Effectiveness in the Analysis of International Environmental Agreements

Gabriela Kütting

This chapter develops an alternative approach to the study of the effectiveness of international environmental agreements that goes beyond the regime-centric traditional approaches. In a critical context the concept of effectiveness will be elaborated to go beyond institutional concerns and existing power relations by adopting an approach that distinguishes between institutional and environmental effectiveness and that locates agreements in their social, structural and ecological context.

A new concept of environmental effectiveness will be introduced that is based on four determinants of effectiveness. These four determinants are science, time, economic and regulatory structures. They highlight the underlying structures and systemic constraints under which international environmental agreements operate. Such an analysis is then embedded into the nature-society dichotomy, which studies the relationship between environmental and social structures.

In the economic, social and political form of organisation that constitutes international society, the state is the foremost regulatory actor. The doctrine of state sovereignty confirms that the state holds a monopoly on legitimacy to enter international negotiations on environmental regulation. In turn, this doctrine is institutionalised in international law and the constitution of the United Nations system. It is not surprising, therefore, that international environmental agreements between states are seen as the principal form of international environmental cooperation. Therefore the state acts as the representative of environmental concerns in the international political economy. This means that international environmental regulation through international agreements is often reduced to the analysis of the various actors in this process rather than focusing on the relationship between environmentally degrading activities and their impact on the social and natural environment.

With the growing requirement that international environmental cooperation be seen to be convincing, this leads one to the question of the effectiveness of such cooperation. The traditional realist and neo-realist view of International Relations (IR) is that cooperation between states as opposed to other forms of cooperation is the only realistic way of focusing on international environmental problems because of the nature of the state system. However, it is argued here that it is not necessarily an adequate or even appropriate way to do so.

The International Relations literature has nominally taken into account effectiveness concerns by locating them within the traditional boundaries of the discipline. This means that it does so with the traditional analytical focus of the discipline, such as scrutinising and trying to explain the behaviour of actors in the international system rather than focusing on the problem giving rise to cooperation and its adequate solution (Haas, Keohane & Levy, 1993; Levy, 1993; Underdal, 1990; Young, 1992). The same also applies to the study of the effectiveness of international environmental agreements. Literature on the subject is mostly limited to how cooperation between states can best be *described* and under what *conditions institutions work*. Effectiveness, as defined by these thinkers, means that an agreement can achieve a change of behaviour in an actor that would not have occurred in the absence of that particular agreement (Young, 1992, p. 161). This understanding does not require that environmental improvement occur as a result of the agreement. However, environmental improvement or healthy ecosystems are a vital precondition for the functioning of social systems as these are dependent on ecosystems for survival through the provision of air, resources, energy and so on.

As such, the effectiveness of an agreement is assessed within conventional IR and regime theory terms, namely in relation to questions of international order and international organisation rather than as an issue of social/economic and environmental compatibility. Regimes are seen as a way to overcome the anarchy of the international system, which, in the realist view, characterises world politics (Morgenthau, 1993 and Bull, 1977), and to create a common code of conduct, or norms. However, though this is the main dimension recognised by the discipline, it is only one dimension of international environmental cooperation. This orthodox focus seems to disregard the paramount purpose of an agreement, namely the publicly declared objective of reducing, halting or otherwise ameliorating environmental degradation. It is this distorting focus of the orthodoxy that places actor behaviour above environmental regulation. The relationship between the international cooperative effort and the environmental problem to be tackled is disregarded by traditional analysis. As Paterson argues:

> Even though it is often accepted that IR as a 'bounded subject matter' may leak through into other areas, there are limits. Thus, for those who have written on global environmental problems, the question 'why have these questions arisen in the first place?' does not usually get asked - this is presumably left to political economists, geographers, or someone else. This question is left to others, as it is supposedly not an 'international' question even though, if you reverse the lens, and start from an environmental perspective, it is of course of supreme importance. Thus the established academic disciplinary boundaries impede our overall understanding of environmental politics, and are one source of why the analyses of environmental regimes are so limited (Paterson, 1995, p. 215).

Although the main question to be asked in this chapter is not why environmental problems have arisen in the first place but, rather, if international environmental agreements can rise to the challenge posed by environmental degradation, the main point made by Paterson is of central importance. The question of how adequately an environmental agreement can and does tackle the environmental problem to be regulated can indeed be deemed to be outside the discipline's boundary, as Paterson argues, because it does not relate to the primary concern of IR, namely to define or find a logic behind the interaction of the various international actors. However, it is argued here that this 'externalisation' (in the sense of placing outside the discipline) is unjustified and needs to be remedied. The debate about 'effectiveness' cannot be limited to institutional recommendations but needs to include an evaluation of the environmental problem that gave rise to the design of an agreement.

In a critical approach suggested by Cox (1981),[1] the boundaries of traditional approaches and the existing structures of international environmental agreements are not taken for granted. Effectiveness needs to relate to the environmental problem giving rise to efforts of international cooperation; it should not simply relate to the existing structure and agency relationships. This author calls this the feasibility-necessity dichotomy. Effectiveness is traditionally defined in terms of institutional feasibility and what can realistically be achieved in terms of actors' aspirations and interests within the particular 'regime'. What is argued here is that what needs to be studied is in fact necessity, not feasibility. This means that more focus should be placed on measures necessary for environmental improvement rather than on measures that are feasible in policy terms but are not helpful for environmental improvement. Existing structures and institutional mechanisms are clearly unsuited to effectively deal with environmental degradation and a concept of effectiveness, both in academic and actual terms, needs to take account of necessities and not just feasibilities.

This chapter is organised in the following way. First, existing concepts of effectiveness are discussed, showing that there is a feasibility-necessity dichotomy. Second, a new concept of effectiveness is put forward that includes both an institutional and an environmental dimension. Third, it is argued that the effectiveness of international environmental agreements is rooted in the social and structural origins of environmental degradation. Therefore, the social, political, economic, technological, scientific and temporal structures surrounding an agreement need to be studied. In addition, the social environment pertaining to the agreement needs to be linked to its 'natural' environment so that the link between that particular environment and society can be studied.

TRADITIONAL APPROACHES TO EFFECTIVENESS

The motivations for studying effectiveness are varied. The institutional school of thought studies effectiveness mainly because it relates to the issue of institutional performance. These writers are mainly regime theorists and as such believe in the vital importance of international institutions/agreements, or regimes, as a form of international order. Effectiveness is used as a measure for regime strength or weakness and thus an effective regime must necessarily be one with a high level of cooperation. The issue of how well this cooperation deals with the problem giving rise to it takes a secondary position. In the environmental effectiveness strand, the concern is with the environmental problem necessitating the agreement. However, these concerns form only part of a wider institutional analysis.

The subordinate role of environmental effectiveness in the existing literature is related to the limited research focus of these texts. Regime theorists assume that a significant and cooperative institution will incite change and change will improve the pre-institutional state of being (Haas, Keohane & Levy, 1993). However, change might, for example, only be an exercise in awareness-raising but for the regime theorist this is a success since it has resulted in an improvement on the pre-regime situation. The standard by which effectiveness is measured is not set in relation to the problem to be dealt with by the regime but by the ability to bring about change. However, this standard is lacking because change by itself is no measurement and does not hold any evidence for a well-functioning institution, nor for environmental improvement. This issue highlights the methodological constraints of regime theoretical assumptions.

Regime theory has an inherent logic which is limited to studying a cooperative effort at the institutional level only. Its shortcomings are founded in its methodology which is static and focused on the unit-level, which means it is reductionist. A reductionist analysis cannot capture the complexity of social and environmental problems and confines itself to modelling institutional set-ups. The variations in regime theoretical approaches are found at the modelling level, not in methodology. The same applies to the effectiveness strand of regime theory.

The effectiveness literature is mainly concerned with the institutional set-up of a regime and how institutional factors influence effectiveness. Although there is a small sector in the literature making a theoretical link between an institution's performance and its effect on environmental degradation, this contribution is not applied in practical terms. For example, for Young, an international environmental agreement

> is effective to the extent that its operation impels actors
> to behave differently than they would if the institution
> did not exist or if some other institutional arrangement
> were put in its place (Young, 1992, p. 161).

Thus, Young defines effectiveness as lying in the performance of the institution. Other determinants such as environmental criteria are seen to be reflected in the performance of the regime/institution and therefore influence effectiveness indirectly. Measurement of effectiveness of an agreement is purely performance-related and therefore dependent on changes in actor behaviour rather than environmental improvement. Thus, no connection between problem-solving adequacy of the regime and its performance is made. This is a classical case of what Cox calls problem-solving theory (Cox, 1981).

Young has established a list of factors that influence the role of regimes (Young, 1992, pp. 176-94). These are divided into exogenous factors relating to the social environment in which the regime operates, and endogenous factors relating to the character of the regime: transparency, (the ease of monitoring or verifying compliance); robustness of social-choice mechanisms; transformation rules, such as ability to adapt to change; capacity of governments to implement provisions; distribution of power such as material inequality between member states; the level of interdependence among member states; and the intellectual order expressed as ideology and the power of ideas. All these factors can help or impede the effectiveness of a regime. An effective regime would need high transparency, high robustness, good transformation rules, high capacity of governments, equal distribution of power, high interdependence and a constant intellectual order. All these factors relate to regime performance.

This regime theoretical example of the study of effectiveness highlights the neglect of environmental concerns. In this light, a summary of the problems with this limited approach to the concept of effectiveness will be made. First, it seems possible to make a distinction between institutional and environmental effectiveness and a good definition needs to incorporate both concepts. However, this distinction is heuristic and for analytical purposes only. Environmental and institutional effectiveness are not two separate concepts but are interconnected and two sides of the same coin. This point will be elaborated below.

Second, there are certain flaws in the definition of institutional effectiveness that need to be improved. The premise favoured by many authors (such as Haas, Keohane & Levy, 1993) that the effectiveness of an agreement can be measured by comparing the achievement of the agreement with a hypothetical state of affairs in the absence of the agreement is not feasible. Moreover, one should not necessarily conclude that an agreement is effective simply because environmental improvement has occurred subsequent to the agreement. This is not feasible as agreements do not exist in a social vacuum. It neglects the consideration of parallel social, environmental, economic, technological and knowledge processes that exist or might exist at the national, regional and indeed global level and influence policy making. Thus, environmental change cannot automatically be attributed to the existence of an international environmental agreement.

Third and importantly, it can clearly be argued that the regime theory school has created a logical inconsistency by attributing major importance to exogenous factors, thus implying that regime effectiveness is not a function of the members' will (Young, 1994). Since regime theory's main tenet is that order is a function of will, it is a contradiction to suppose that effective order is not. Therefore there is a fundamental contradiction between one of the principles of regime theory and regime theoretical analysis of effectiveness.

Fourth, it remains unclear why traditionally the effectiveness of an agreement is only related to environmental indicators in a very few cases. The difficulty associated with the task of relating environmental indicators to the effectiveness of international environmental agreements cannot be the cause of this as some institutional indicators used in traditional analysis (for example, the attribution of environmental change to the regulatory efforts of an agreement) are far more difficult to measure.

Finally, there is a vagueness surrounding these existing definitions as they all operate outside a time frame, yet effectiveness should be an issue very much associated with time constraints. Environmental change is usually irreversible and thus it seems imperative that a process of degradation has to be halted as early as possible to minimise irreversibility. Therefore it can be argued that the study of effectiveness of international environmental agreements necessitates a meta-theoretical approach for which the regime theory method is ill-equipped.

INTRODUCING THE CONCEPT OF ENVIRONMENTAL EFFECTIVENESS

The discussion on the underlying foci in the study of effectiveness has demonstrated quite clearly that there has to be at least some reference to the environmental problem giving rise to the agreement. Therefore it seems obvious that regime theory can offer neither a useful nor a workable definition of effectiveness. However, what we can learn from regime theory is that in order to have effective environmental agreements we need to have functioning institutions. Since institutions are what regime theory focuses on, the experience of this research has to be valued but treated with caution due to its reductionism and its treatment of agreements as closed systems. It has to be placed in a different analytical framework.

So far, effectiveness has basically been defined as a well-working institution whose performance can achieve change in its members' behaviour. This emphasis has led to a prioritising of the analysis of agreement implementation rather than agreement formation. In turn, this has meant that effectiveness has come to refer to changes actually achieved by the agreement in question - whether good or bad. In contrast, the prescription of remedies in the first place has been neglected in analytical terms which means that environmental amelioration as a concept does not form a major part of analysis. This latter notion is what is developed here. The distinction between institutional and environmental effectiveness is made most obvious in this context.

Institutional effectiveness is mostly concerned with the performance of the institution in question while environmental effectiveness as a concept makes the eradication or prevention of environmental degradation its priority. Neither of the two types of effectiveness are exclusive in their approach and therefore the distinction between the two can only be made heuristically. In addition, institutional effectiveness is reductionist, pragmatic and mostly concerned with feasibility. Environmental effectiveness, on the other hand, takes a holistic approach and is idealistic in the sense that it concentrates on necessity rather than feasibility. In this sense, institutional effectiveness is used to describe existing approaches. However, environmental effectiveness as a concept overcomes the methodological limitation of institutional effectiveness by incorporating its concerns in a wider framework that makes the institution in question part of a web of social, political, economic and environmental relations. It studies the effectiveness of international environmental agreements not from a traditional problem-solving theory aspect but takes recourse to a holistic analysis in a critical context.

Institutional effectiveness

The concept of institutional effectiveness can be largely equated with the definition of effectiveness as put forward by the regime analysts. However, this needs to be qualified by adding that an agreement can only be called institutionally effective when its administrative framework and working process are actually geared towards solving the problem in response to which it was formulated. Leaning heavily on regime analytical approaches, the following conditions apply to an institutionally effective agreement:

- Participation includes all those affected by the problem, both polluter and polluted.
- The framework includes provisions for increasing knowledge on the environmental problem and is able to incorporate the knowledge created.[2]
- Linkage of environmental policy to other policy issues (Susskind, 1994, p. 7).
- A high degree of good will.[3]
- Achievement of the institutional goals (Wettestad & Andresen, 1991, p. 2).

The exogenous-endogenous factor distinction has been directed toward determining the *sources* of effectiveness. It is based on the assumption that there are sources of effectiveness that, once tapped, will lead to increased order in the international system. The above list refers to the *conditions* of effectiveness: the achievement of institutional effectiveness depends on certain conditions that may not be met in every agreement. On the other hand, a focus on sources of effectiveness (robustness, interdependence, power distribution and so on) implies that effectiveness can be achieved through the right constellation of factors and is therefore a function of will. Therefore, the argument put forward here is that the effectiveness of an agreement is not dependent on sources but rather on conditions that have to be met.

However, institutional effectiveness is a reductionist concept and is methodologically limited unless placed in a wider systemic and structural context of which it is only part. The concept of environmental effectiveness treats agreements not as closed systems but as interconnected parts in a much wider social, political, economic and environmental web, which needs to be studied in order to understand the complexities underlying environmental degradation.

Environmental effectiveness

The environmental effectiveness of an international environmental agreement refers to the degree to which the degrading or polluting processes and consequences are arrested or reversed. On the surface, this definition does not seem to be any different from Haas, Keohane & Levy's definition stipulating that an effective agreement/institution is one resulting in change as compared to the pre-agreement state. However, there are vital differences.

First, it is not necessary that the reversal of environmental degradation is directly attributable to the agreement in question. If the agreement set in motion other social processes that lead to an arrest or reversal of degradation, then the agreement has been indirectly effective. Second, this definition does not only take change in actor behaviour as its basis but a clearing up, or solution, of the problem in question – this point is not included by Haas, Keohane & Levy or other authors (1993). Third, the new definition places the agreement in a wider context, both in the societal and environmental sense, by studying it with a particular emphasis on the four determinants of effectiveness: science, time, regulatory and economic structures. Existing approaches do not go beyond institutional concerns.

The four determinants

It is considered necessary to place an agreement in the context of these four determinants in order to define its relationship with society and the environment. A definition of this relationship is vital for determining an agreement's effectiveness.

(a) *Regulatory structures*

Regulatory structures evolve very much out of the social, political and economic context in which the agreement operates. Therefore, they reflect institutional concerns and are geared towards bringing member states together on the administrative level. However, as a result, environmental concerns are sidelined because they will complicate the institutional consensus. Bureaucracy can only reflect societal but not environmental structures and therefore cannot relate to the issue giving rise to its existence.

Since the provisions of international environmental agreements are the result of the respective policy-making process, it is necessary to study this process in order to understand its social significance. This involves a study of both structure and agency. Regime theoretical analysis is limited because its methodology concentrates on agency. The change over time and the evolution of international environmental agreements can only be analysed and explained by looking at the changes in the policy-making process. This necessitates a political economy approach that goes beyond the limited focus of regime theoretical approaches. Such an approach also needs to be located temporally and spatially in order not to be in danger of being static. The policy-making process is not isolated from the general evolution of society and is therefore an indicator of social change that needs to be analysed in a political economy approach.

With a holistic view, the policy-making process and its results can be placed in a historical perspective as well as seen in conjunction with other regional or global trends. This will demonstrate that the provisions negotiated in international environmental agreements are not as fortuitous or actor-driven as they seem but also depend to a large extent on wider processes of which they are only part, and are rooted in particular social and structural origins. Some of these processes such as environmental degradation, are not consciously regarded as being part of wider social processes. This means that they are external to and excluded from analysis. A holistic view can take this into consideration and rectify the shortcomings.

The meaning of this phenomenon has wide-reaching consequences. It means that in policy making, the process is not directed intentionally but is a composition of a myriad of unit-levels which are not co-ordinated to form a macro-level but the outcome is one of coincidence, without deliberate direction. Conventional theories are able to explain actor behaviour because they work on the same limited principle as the actor/policy maker. These theories use as their basis the same type of information available to the policy maker and thus also operate on the micro level only.

Beck has taken up this phenomenon of the society fragmented in its knowledge structures and has highlighted the major feature of what he terms the risk society (Beck 1986 and especially 1995). He distinguishes between risks that were created in active and conscious decisions, and which can thus be controlled, and those risks that have avoided social control mechanisms. These take place on two levels. On the one hand, the institutions and norms developed and refined by industrial society (risk calculation, insurance, preventative cure and so on) fail in the light of controversial technology use such as nuclear, genetic, industrial and chemical processes. These are not accepted as risks by insurance companies and remain uninsured whereas they have to be accepted as negligible risks by society. On the other hand, decisions made on the national or firm level involve the risks which affect all members of the world society. As Beck argues,

> Mit dem ökologischen Diskurs wird das Ende der 'Außenpolitik', das Ende der 'inneren Angelegenheiten eines anderen Landes', das Ende des Nationalstaates alltäglich erfahrbar (Beck, 1995, p. 16). [The ecological discourse makes us experience the end of 'foreign policy', the end of 'sovereignty' and the end of the nation state every day. *Translated by author*]

As environmental damage is taken out of the temporal and spatial dimension, the 'polluter pays' principle is applied in an opaque way and compensation is not possible due to the uninsurability of the risks involved (Beck, 1995, p. 16).

The question that follows Beck's argument is how national policy makers with their limited expertise and accountability can legitimately deal with the problems of what Beck terms the ecological world risk society described above. They do not understand, and are not required by their profession to understand, the structures of the world risk society as well as the way it has evolved and is evolving. At the same time, they still have to decide how it is regulated with their narrow concept of institutional feasibility.

Beck sees as a solution the increased awareness of this problem demonstrated by the rise in international environmental agreements but also by action taken by non-industrial actors as well as by the so-called green industry, which evolved as an antithesis to the risk industries (Beck, 1995, p. 19).

Beck's argument that the proliferation of international environmental agreements in the past 20-30 years witnesses a development that counteracts the evolution of the risk society, however, is not valid for several reasons. First, these agreements are not related to processes outside the negotiating process and are thus unaware of their role as ascribed by Beck. Second, the participants in the negotiating process are not aware of the structures of the risk society because social structures are outside their expertise and horizon. Third, and most importantly, the assumption that international agreements control environmental risks is incorrect as the study of effectiveness demonstrates.

This rather detailed discussion of the risk society argument is necessary in order to highlight its temporal and spatial dimension. The trend of increased specialisation and thus fragmentation of social processes is a global one and not only specific to national policy makers in environmental negotiations. Part of it can be explained with bureaucratic theories, bureaucracies taking on a life of their own and justifying their existence by constantly inflating themselves. However, the main problem is that bureaucratic institutions, and especially policy makers, get their motivation from the inside and thus their interests are guided from the inside as well. Since environmental degradation is an externality, it is outside bureaucratic processes.

This does not mean that bureaucratic processes are hermetically sealed against outside influences. So, factors such as public opinion, scientific expertise from non-governmental sources, and so on can still influence the policy-making process, although filtered through bureaucratic channels. This means that evolution and change of processes are necessarily long-term phenomena. In contrast, environmental degradation is a problem that necessitates swift reactions. Hence, there is an incompatibility.

(b) *Economic structures*

Economic structures face a similar problem. They determine social organisation in general and therefore condition the way environmental issues are perceived by society. Therefore the effectiveness of international environmental regulation is dependent on a connection between economy and environmental regulation.

Economic structures are a very clear determinant of effectiveness because of the nature of social organisation. Economic organisation determines the form and shape of social organisation in general (Giddens, 1990 and Beck, 1986/95). It also determines the way the environment is perceived by society. For example, the view that green technology can overcome environmental degradation and the connected idea of sustainable development are based on notions of capitalist ideology that presume that growth can continue indefinitely. The link between society and its environment is not fully understood.

At the institutional level, this means that regulatory structures reflect this economic determinism and negotiations are guided very much by economic considerations and feasibilities as the sections on regulatory structures above as well as on science and time below suggest. The sheer fortuity of an environmentally friendly but economically viable alternative to existing processes determines institutional success or failure. Again, this point demonstrates the poverty of regime theoretical analysis. In terms of environmental effectiveness, economic structures provide the framework through which environmental degradation occurs and through which it can be avoided.

At this point it is timely to discuss briefly Cox's writings on social forces and world order. Cox's main concerns are production and power but his critical approach nevertheless has some applicability for the argument of this section of the chapter. Cox uses his critical theory to study the changing role of the forces of production in world order over time and identifies changes in political world order that go hand in hand with changes in the configuration of social forces, particularly forces of production (Cox, 1987). Cox's main contribution to the study of IR is the introduction of a long-term Braudelian study of global political economy. Therefore its outlook and concerns are related to world history and the history of social change. Although the concern of this chapter is social change relating to environmental degradation rather than social change *per se*, a link can be made between Cox and this argument in terms of the four determinants of effectiveness, particularly economic structures. This link is based on the premise that 'production creates the material basis for all forms of social existence, and the ways in which human efforts are combined in productive processes affect all other aspects of social life, including the polity' (Cox, 1987, p. 1).

Power and production influence each other. This holds for the four determinants of effectiveness: Regulatory mechanisms, economic structures and the political economy of science are determined by productive processes and their dominant role as social forces. Even the temporal organisation of society is dominated by the consideration of production. However, this is exactly where the issue of effectiveness becomes important. Because social organisation is so intertwined with production, improvement of the degrading consequences of the productive process cannot be incorporated in a social framework that is geared towards facilitating production. Cox himself recognises this:

> The global economy, activated by profit maximisation, has not been constrained to moderate its destructive ecological effects. There is no authoritative regulator, so far only several interventions through the inter-state system to achieve agreement on avoidance of specific noxious practices (Cox, 1996, p. 516).

However, Cox sees the main analytical challenge in this to be the effect on the state system and sovereignty rather than the challenge to overcome environmental degradation. This is the limitation of his approach because it demonstrates that Cox's critical approach still operates within the orthodox concerns of the IR discipline; it is still concerned with actors in the international arena. It does not offer an opportunity for incorporating the concern of effectiveness in relation to international environmental agreements as is made abundantly clear:

> The biosphere has its own automatic enforcers, for
> instance, in the consequences of global warming; but
> who will negotiate on behalf of the biosphere? That
> must be one of the questions overshadowing future
> multilateralism (Cox, 1996, p. 517).

The critical approach taken above implies that the question
overshadowing environmental degradation is related to who will
negotiate in what ways rather than how environmental degradation
can be remedied given the mutual constitution of power and
production.

This section has demonstrated that economic organisation and
environmental degradation are inextricably linked since it is the
mass production and consumption of goods that leads to
environmental degradation. However, the production structure and
productive forces are the determinants of social organisation and
also determine the way society values and counteracts
environmental degradation. Therefore it can be argued that the
determinant of effectiveness relating to economic structures in
many ways incorporates the other three determinants.

(c) *Science*

Another determinant of environmental effectiveness can be found in
the social role of science. The IR literature on international
environmental agreements limits the role of science to the input of
science into agreement-making (Haas, 1990; Sjöstedt, 1993;
Susskind, 1994). However, this limitation does not do justice to the
influence of science in all social spheres. Science determines the
everyday life of society and has an all-pervasive influence on it
since lifestyle in late 20th century is largely based on rational
scientific logic. In western culture, science has become an ideology
that provides the core values of society, even replacing the role of
religion (Midgley, 1992). In this context, science is taken to mean
the activity, and its results, carried out by a professional group of
people in universities and other research institutions (the scientists
trying to find laws and correlations in their study of phenomena

occurring in the physical environment by simulating these in a laboratory environment.[4] This method of analysis isolates the environmental problem from its ecosystemic context, and this practice may lead to complications when applied. As Jasanoff argues, 'scientific inquiry, contrary to expectation, does not always lead to the same explanation for the same observed phenomenon' (Jasanoff, 1995, p. 147). The main issue at stake is that scientific activity is not an isolated social activity but one that is interconnected with other social processes.

Jasanoff takes up this point in a different manner. She quotes Brooks to make a distinction between *science in policy* and *policy for science*:

> The first is concerned with matters that are basically political or administrative but are significantly dependent on technical factors - such as the nuclear test ban, disarmament policy, or the use of science in international relations. The second is concerned with the development of policies for the management and support of the national scientific enterprise and with the selection and evaluation of substantive scientific programs.[5]

This distinction is not mutually exclusive. Policies for science necessarily need to be made with the advice of scientists and science in policy obviously needs to be conducted in such a way as to fit in with administrative and regulatory structures. This makes the distinction more confusing than helpful at times but also shows that science is not a separate domain but just another sector of social organisation. Likewise, the distinction of scientific expertise as determined by the background of the scientist (environmentalist or industry representative) shows that science and economics are intrinsically linked (Rodricks, 1992). Jasanoff's work also shows that science is preoccupied with institutional effectiveness rather than environmental degradation. As she argues in relation to risk in policy making:

> Judgements about risk inevitably incorporate tacit
> understandings concerning agency and responsibility,
> and these are by no means universally shared even in
> similarly situated western societies. Against this
> background it makes little sense to regulate public
> demands, and claims to superior expertise.
> Environmental regulation calls for a more open-ended
> process, with multiple access points for dissenting views
> (1997, p. 13).

In agreement-making terms this means that scientific advice is
definitely not an objective source of knowledge that needs to be
tapped in order to make good policies and agreements as implied in
the IR literature. Therefore the notion of independent expert advice
providing guidelines for effective policy making is misleading.
Rather, it is more useful to refer to a political economy of science
that underlies agreement-making and influences its outcomes.

(d) *Time*

The fourth determinant of environmental effectiveness relates to
notions of time. Time is not just a measurement according to which
we plan our schedules, be it short-term, long-term, day-to-day and
so on, but can be institutional or social, cyclical or linear, perceived
or measured, according to the focus of analysis. Time in its various
forms is such an exceedingly important issue not only because of
the irreversibility of environmental degradation but also because it
dominates every society and individual's life as all organisation is
ultimately based on time issues. Therefore, analysing the
significance of perceptions of time has a central place in policy
terms due to the problem of the irreversible loss associated with
environmental degradation. However, this aspect of urgency related
to time is not translated in policy terms.

Environment and society are treated as separate phenomena by
social scientists. The environment is subjected to human
perceptions of time in the way it is treated and analysed. However,
it operates under non-human time frames. Therefore, it is in a
special position compared to social problems that are studied. This

special status and its consequences will be briefly touched upon here. Adam describes this phenomenon:

> How can [we] make sense of the different definitions, approaches and proposals for solutions [to environmental degradation] when [they] have been established on the irreducible distinction between nature and culture, the natural and the symbolic environment, evolution and history, when the environment as a subject matter so clearly falls outside [this] traditional bounded domain? (1994, p. 93).

As suggested by the above statement, in the social sciences, the realisation that the environment is not a 'given' within which society operates is relatively new. This lack of acknowledgement explains the literary vacuum on the study of time and the environment. It also demonstrates the need for an interdisciplinary approach to it.

Relating to the effectiveness of international environmental agreements, there are two primary time issues that are important. First, on an institutional as well as environmental level, time frames of international environmental agreements need to reflect environmental necessities. Institutionally this means that the administrative process from the formation of an agreement to its implementation and the time frames imposed by the agreement need to reflect the urgency and irreversibility of the environmental problem to be regulated.

An agreement has to go through a process of negotiation, signature, ratification, entry into force and then implementation. The text of the agreement may or may not prescribe targets in what time period the provisions of the agreement have to be achieved.

Administrative time frames relate to the institutional level and should thus be an issue or condition of institutional effectiveness. However, the concept of time has so far been neglected by the institutional effectiveness school of thought. This neglect applies to both time frames as a measure of institutional effectiveness and to the connection between time frames and environmental amelioration. Thus administrative time frames need to be incorporated into institutional effectiveness analysis, which is a part of environmental effectiveness analysis.

The other primary time issue relates to rhythmicity and how social activity can disrupt environmental rhythms. Social activity impacting on environmental processes can primarily be found in the industrial sector, particularly mechanical and technological processes.

Environmental and technological processes do not share the same underlying principles according to which they evolve or function. Environmental processes are highly interactive, rhythmic, cyclical and 'renewable' (Daly, 1992, Lovelock, 1988). Technological processes, on the other hand, are extremely linear (this means non-renewable after their life span expires, and also producing waste). Although some social technological processes can be interactive and rhythmic, technological products used in the production process certainly are not. A machine's life is mechanical and non-renewable, this means functions according to Newtonian principles. This means that a technology-centred economy is based 'on the principles of decontextualisation, isolation, fragmentation, reversible motion, abstract time and space, predictability, and objectivity, maxims that stand opposed to organic principles such as embedded contextuality, networked interconnectedness, irreversible change and contingency' (Adam & Kütting, 1995, p. 243).

Therefore, two systems have to cohabit, namely the environment and industrial society, that are based on opposing principles. Industrial society interferes with the rhythm of nature and thus disturbs its careful balance by disturbing entropic processes. This phenomenon can be summarised by distinguishing between mechanism and organism. Mechanical systems draw on Newtonian concepts and assumptions, which are characterised by the following: decontextualisation and abstraction; interacting parts making up self-contained systems; whole and parts are isolatable in an absolute way; parts are interchangeable and thus replaceable; focus is on function, not relations; emphasis is on efficiency and predictability; energy is consumed but not given back to the system; and rhythms are abstract, reversible and invariable. Organic systems, on the other hand, can be described as: contextual; embedded; networked connectivity; open; interactive systems of exchange; part-whole implication with no isolatable parts; parts cannot be exchanged without affecting the whole; focus on relations; interaction and exchange; emphasis on processes of life and death; use of energy resulting in growth and life and rhythms are cyclical, evolutionary and irreversible. By removing resources from the environment and returning them in the form of waste only, mechanical systems, or industrial society, endanger the functioning of these cyclical, regenerative phenomena, which results in environmental degradation, which then affects society again. Therefore there is a fundamental incompatibilty in temporal terms between environment and contemporary society.

The discussion of the four determinants demonstrates that existing IR approaches are not really well suited to study the environment since they operate on the premise that social systems are the only, or at least major, systems in existence. However, it is not sufficient to incorporate the environment as a new subject or actor in existing modes of analysis because clearly this is not what it is. Rather, it is a system of which all other (human/social) systems form part. Thus actor-centric analysis will not be very fruitful and a systemic/structural approach must be chosen that goes beyond the sole consideration of social structures and forces. However, such an approach needs to include the environment not as another system but as a system of which all other systems are part.

This is clearly a difficult task since it goes against prevailing concepts. A study that attributes causal liability to the functioning of environmental systems must therefore displace the exclusive preoccupation with institutional factors, and instead concentrate on environmental effectiveness, or compatibility.

THE LINK BETWEEN ENVIRONMENT AND SOCIETY

The environment-society link will show systematically why society and social actions cannot be studied without understanding the natural environment in which they operate. The environment has only recently started to be the subject of social scientific analysis. The use of natural resources for the production process and the abuse of the natural environment as a 'waste bin' or sink had not

been seen as a problem until the 1960s and not been taken seriously
as a problem until the 1980s. This is hardly surprising since, after
all, social scientists are concerned with the study of society, not the
study of the environment.

The relationship between nature and society can be studied
from an historical angle and ecological economists such as Daly
trace the origins of environmental degradation to the beginnings of
the industrial revolution and the resulting unbalancing of the state
of entropy (Daly, 1992). Others, on the other hand, find that
humankind irreversibly changed the face of the globe even at the
hunting and gathering state of human evolution by converting land
for agricultural purposes, hunting to extinction of animals and
overexploitation in general (Ponting, 1991). It is undeniable that the
phenomena   of   pollution   and   degradation   existed   before
industrialisation but these were rather localised and did not affect
the general equilibrium of the global ecosystem. Therefore the
industrial revolution with the consequential changes in economic,
political and social organisation has to be seen as the main social
origin of environmental degradation in historical terms.

Redclift and Woodgate give a summary of the different
interpretations that exist on the relationship between nature and
society (Redclift & Woodgate, 1994). The phenomenon of co-
evolution sees society as part of the physical processes. Co-
evolution can be described as an interactive process between society
and nature while nature changes through evolution and society
changes through processes of structuration. With historical changes,
society has increasingly taken over the role of nature as changes in
agriculture demonstrate. Aspects of nature have been integrated in
the production process and placed under scientific control. Thus
increasingly large parts of nature are under the control of society.

The methodological constraints of actor-centric regime
theoretical approaches cannot capture the complexity of structures
that underlie any attempt to regulate environmental degradation.
Temporal, economic and scientific structures can affect the study of
society and the environment and an understanding of these concepts
is vital for systematic analysis of problems related to environmental
degradation.

It can be established that negotiating processes are part of general social developments. More specifically, processes are not conscious driven phenomena, but they happen rather as fortuitous consequences of a bulk of decisions taken on the micro-level. This demonstrates that national policies are not only formed on the basis of power and economic interest of one single actor but with reference to various sub-actors, political culture, environmental and ideological values and so on that need to be accommodated.

This analysis explains why only institutional and not environmental effectiveness is an issue in international environmental agreement-making. However, it gives no hope for the inclusion of environmental effectiveness as a concept in policy making, except as an issue highlighted by more or less integrated environmental NGOs, which operate on the margins of decision making.

In academic and in policy-making terms, there is no movement towards a holistic rather than socio-centric form of policy making that transcends the environment-society divide. Academic disciplines such as politics, sociology, economics and IR are anthropocentric by definition. Even when phenomena focusing on the whole planet such as time are the focus of analysis, they are still considered from a point of view that is focused on social aspects. The example of the effectiveness debate demonstrates this very clearly. Moreover, the treatment of the issue of time illustrates the shortcomings and limitations of this approach more clearly than other debates. However, the dominance of the regime theory/institutional bargaining school of thought shows us how embedded conventional behavioural and policy analysis is and how the policy makers themselves are driven by the notions focused on in this school of thought. Therefore it cannot be expected that fundamental changes will occur in the policy-making process in the near future.

It can be concluded from this that the environmental effectiveness of international environmental agreements can only be achieved if the policies proposed transcend approaches that are based on human rhythmicity and reintegrate social (including political and economic) rhythms to make them part of natural rhythms rather than dominate them. This implies a changed rhythmicity. International environmental policy making is dominated by social concerns and perspectives and suffers from such a degree of institutionalisation that it cannot seriously be expected to transcend these values. This means that by definition environmental effectiveness will not be achieved through an institutionalised agreement-making system, this means international environmental agreements cannot be effective by their very nature.

Two issues create a connection between the effectiveness debate and the contextualising of the concept of effectiveness. First, it has become obvious that the narrow focus on institutional effectiveness by regime theorists cannot adequately capture the meaning of effectiveness in a holistic context or in the context of a particular case study. Second, this leads to the realisation that effectiveness, or indeed environmental agreements, cannot be studied in a rigid, limited methodological context but have to be placed in a wider social, political and economic context in order to be understood and analysed in a meaningful way.

Temporal, economic and scientific structures form the basis of any analysis of the relationship between society and the environment and an understanding of these concepts is vital for understanding problems related to environmental degradation. The social mechanisms directed at regulating or alleviating environmental degradation work according to the rhythm of industrial society and are thus unable to transcend this limitation unless a reform of social and economic organisation takes place. International environmental agreements demonstrate this clearly. They are so dominated by considerations about different societies' interest, values, motivations, political systems and so on that they cannot incorporate issues such as different rhythmicities and even focusing on the problem at stake. Therefore, they cannot offer a viable solution to the problem of environmental degradation.

International environmental agreements cannot be effective by their very nature because they operate under such structural constraints. A fundamental restructuring of social organisation in order to take account of historically rooted environmental ignorance is necessary to overcome this problem. Therefore, environmental effectiveness is a concept that needs to be incorporated into social and economic analysis.

CONCLUSIONS

The mainstream IR literature ignores the complexity of environment-society relationships by concentrating on institution-building and aspects of international cooperation. However, the environmental problématique shows that a fundamental rethinking has to take place in order to take account of a field of study that has been completely externalised. Overall, it is a structural level on which all societies are dependent. Therefore it is a fatal omission to externalise it and to study its institutional aspect only.

The traditional concern of study in IR has been the behaviour and/or relationships between states and other international actors. However, the case has been made that this concern is not far-reaching enough. It is not just the actions and the behaviour of international actors trying to cooperate on an issue that needs to be studied but even more so how they will do it. It is not sufficient to try and explain the behaviour of actors and their motivations if this does not lead to analysis of how environmental problems can be dealt with more effectively. This is true of social problems but becomes paramount in relation to environmental degradation.

Another issue associated with this point is the methodological limitation of traditional IR case studies. Reductionist analysis serves to explain immediate connections between motivation and action but cannot capture the full web of complexities in which a particular agreement-making process or another form of cooperation is located. Trade-offs are made between different subject matters, so even in institutional terms such explanations are too limited. However, in its most obvious form this parsimonious approach with its selection of only a few key variables neglects the social and structural context in which a particular form of cooperation takes place. This has also been noted by Paterson in a slightly different context (Paterson, 1996). He argues that realist, neorealist and neoliberal institutional approaches can only partially explain actor behaviour in the case of global warming but historical materialist political economy approaches are substantially more successful in explaining negotiating positions than the traditional IR approaches. However, Paterson limits himself to explaining positions rather than linking the issue to environmental degradation, therefore still operating within traditional IR concerns.

These findings are also applicable to non-environmental areas of IR. They are valid in other areas where the literature adopts a scientific analysis based on rational criteria which do not address the underlying issues. Realist, neorealist and regime theoretical approaches to poverty, human rights, food aid, the North-South divide, development issues in general or property rights are a case in point.

These issues are analysed in terms that are not related to the problem giving rise to international cooperation or consultation. Therefore the problem itself is externalised from the analysis, which means that ethics, moral values and notions of responsibility or equality and equity are not addressed. In addition, the social and structural origins of a problem can be ignored in traditional actor-centric analysis which relates directly to the aforementioned point. This means that, apart from the environment-society divide, the same criticisms made in relation to the effectiveness of international environmental agreements also apply to the above issues of poverty, human rights and so on.

NOTES

1. Cox distinguishes between problem-solving theory and critical theory. Problem-solving theory reproduces prevailing power and social relationships with the general aim of problem solving being 'to make these relationships and institutions work smoothly by dealing effectively with particular sources of trouble' (p. 129). The name problem-solving theory is confusing because it gives the impression that the main concern is the resolution of a problem. However, this is not the case. Critical theory, on the other hand, stands apart from prevailing orders and structures by not taking institutions and social/power relationships for granted. Rather, it is directed 'towards an appraisal of the very framework for action, or problematic, which problem-solving theory accepts as its parameters' (p. 129).

2. These two points are distilled from legal arguments in different contexts.

3. This point has been adapted from the points made by P. Haas, R. Keohane & M. Levy (1993), and O. Young (1992).

4. This point is an adaptation of Kuhn (1982), p. 75.

5. Brooks (1990), p. 76, in: S. Jasanoff, (1997), p. 5.

6. Boehmer-Christiansen, for example, distinguishes between 10 functions of scientific expertise, many of which emphasise the politicised nature of science (Boehmer-Christiansen, 1995), pp. 195-203.

REFERENCES

B. Adam, *Time and Social Theory* (Cambridge: Polity Press, 1990).

B. Adam, 'Running out of time; global crisis and human engagement', in Redclift, M. & Benton, T. (eds), *Social Theory and the Global Environment* (London: Routledge, 1994).

B. Adam, & G. Kütting, 'Time to Reconceptualise 'Green Technology' in the Context of Globalisation and International Relations', *Innovation, The European Journal of Social Sciences*, 8 (3), (1995) 243-259.

E. Altvater, 'Die Oekologie der Neuen Welt(un)Ordnung', *Nord-Süd aktuell*, 7 (1), (1993) 72-84.

B. Barnes, & D. Edge, *Science in Context* (Milton Keyes: Open University Press, 1982).

U. Beck, 'Weltrisikogesellschaft: zur politischen Dynamik globaler Gefahren', *Internationale Politik*, No. 8, (1995) 13-20.

U. Beck, *Risikogesellschaft; auf dem Weg in eine andere Moderne* (Frankfurt am Main: Suhrkamp, 1986).

T. Bernauer, 'The Effect of International Environmental Institutions: How we Might Learn More', *International Organization*, 49 (2), (1995) 351-377.

S. Boehmer-Christiansen, 'Reflections on Scientific Advice and EC Transboundary Pollution Policy', *Science and Public Policy*, 22 (3), (1995) 195-203.

H. Brooks, 'The Scientific Adviser', in R. Gilpin & C. Wright (eds), *Scientists and National Policy Making* (New York: Columbia University Press, 1990).

H. Bull, *The Anarchical Society* (London: Macmillan, 1977).

R. Carson, *Silent Spring* (London: Penguin, 1958).

K. Conca, M. Alberty, & G. Dabelko, (eds), *Green Planet Blues* (Boulder, Colorado: Westview, 1995).

R. Cox, *Approaches to World Order* (Cambridge: Cambridge University Press, 1996).

R. Cox, *Production, Power and World Order; Social Forces in the Making of History* (New York: Columbia University Press, 1987).

R. Cox, 'Social forces, states and world orders: beyond International Relations theory', *Millennium: Journal of International Studies*, 10 (2), (1981) 126-151.

H. Daly, *Steady-state economics*, 2nd ed. (London: Earthscan, 1992).

P. Dickens, *Society and Nature - towards a green social theory* (London: Harvester Wheatsheaf, 1992).

A. Dobson, *Green Political Thought*, 2nd ed. (London: Routledge, 1995).

H. French, 'Wirksame Gestaltung von Umweltschutzabkommen', *Spektrum der Wissenschaft*, 16 (February), (1995) 62-66.

A. Giddens, *The Consequences of Modernity* (Stanford: Stanford University Press, 1990).

S. Gill, and J. Mittelman, (eds), *Innovation and transformation in International Studies* (Cambridge: Cambridge University Press, 1997).

P. Haas, *Saving the Mediterranean* (New York: Columbia University Press, 1990).

P. Haas, R. Keohane, & M. Levy, (eds), *Institutions for the earth: sources of effective international environmental protection* (Cambridge/Mass.: MIT Press, 1993).

G. Hardin, 'The tragedy of the commons', *Science*, Vol. 162 (1968).

W. Hein, 'Die Neue Weltordnung und das Ende des Nationalstaats', *Nord-Süd aktuell*, 7 (1), (1993) 50-59.

M. Hollis, & S. Smith, *Explaining and Understanding International Relations* (Oxford: Clarendon Press, 1990).

S. Jasanoff, *The Fifth Branch, Scientists as Policy Makers* (London, Harvard University Press, 1990)

S. Jasanoff, 'Skinning Scientific Cats', in K. Conca, M. Alberty & G. Dabelko (eds), *Green Planet Blues* (Westview Press: Boulder/Colorado, 1995).

S. Jasanoff, *Regulating Environmental Risks*, Environmental futures lecture series, Cambridge, 25 February (1997).

T.S. Kuhn, 'Normal Measurement and Reasonable Agreement', in B. Barnes & D. Edge (eds), *Science in Context* (The Open University Press, Milton Keynes, 1982).

M. Levy, 'Political Science and the Question of Effectiveness of International Environmental Institutions', *International Challenges*, 13, (1993) 17-35.

J. Lovelock, *The ages of Gaia*, 2nd ed. (Oxford: Oxford University Press, 1988).

M. Midgley, *Science as Salvation* (London: Routledge, 1992).

H. Morgenthau, *Politics Among Nations* (New York: McGraw-Hill, 1993).

M. Paterson, *Global Warming and Global Politics* (London: Routledge, 1996).

M. Paterson, 'Radicalising regimes? Ecology and the Critique of IR Theory', in J. MacMillan, & A. Linklater, (eds) *Boundaries in Question* (London: Pinter Publishers, 1995).

C. Ponting, *A Green History of the World* (London: Penguin, 1991).

R.D. Putnam, 'Diplomacy and Domestic Politics: The Logic of Two-level Games', *International Organization*, 42 (3), (1988) 427-460.

M. Redclift, *Wasted - Counting the Costs of Global Consumption* (London: Earthscan, 1996).

M. Redclift, & G. Woodgate, 'Sociology and the environment. Discordant discourse?' in M. Redclift & T. Benton (eds), *Social Theory and the Global Environment* (London: Routledge, 1994) pp. 51-66.

P. Reynolds, *An Introduction to International Relations* (New York: Longman, 1980).

J.U. Rodricks, *Calculated Risks* (Cambridge: Cambridge University Press, 1992).

J. Rosenau, & E. Czempiel, (eds), *Governance Without Government: Order and Change in World Politics* (Cambridge: Cambridge University Press, 1992).

P. Sand, et al (ed.), *The Effectiveness of International Environmental Agreements* (Cambridge: Grotius Publications Ltd, 1992).

E. Schumacher, *Small is Beautiful* (London: Vintage, 1973).

G. Sjöstedt, (ed.), *International Environmental Negotiation* (London: Sage, 1993).

J. Skjaerseth, 'The Effectiveness of the Mediterranean Action Plan', *International Environmental Affairs*, 6 (4), (1994) 313-334.

R. Stubbs, & G. Underhill, (eds), *Political Economy and the Changing Global Order* (London: MacMillan, 1994).

L. Susskind, *Environmental Diplomacy; Negotiating More Effective Global Agreements* (Oxford: Oxford University Press, 1994).

C. Thomas, (ed.), *Rio: Unravelling the Consequences* (London: Frank Cass, 1994)

A. Underdal, 'The Concept of Regime Effectiveness', *Cooperation and Conflict*, 27 (3), (1992) 227-240.

E.U. von Weizsäcker, *Earth Politics* (London: Zed Books, 1994).

I. Wallerstein, *The Politics of the World-Economy* (Cambridge: Cambridge University Press, 1984).

J. Wettestad, &. S. Andresen, *The Effectiveness of International Resource Cooperation: Some Preliminary Findings* (Fridtjof Nansen Institute: Lysaker, 1994).

O. Young, *International Governance, Protecting the Environment in a Stateless Society* (London: Cornell University Press, 1994).

O. Young, 'The Effectiveness of International Institutions: Hard Cases and Critical Variables', in Rosenau, J. & Czempiel, E. (eds), *Governance without Government: Order and Change in World Politics* (Cambridge: Cambridge University Press, 1992).

Part III: Issues in Negotiations

4 Dynamics of Environmental Negotiations

Ho-Won Jeong

Multilateral negotiations are designed to establish a set of goals and formal mechanisms for implementing policy options. The content, types and nature of agreements reflect interactive processes of combining diverse interests and values. International environmental negotiation constitutes a multi-faceted process: parties should agree on a set of facts associated with problems, recognise different perceptions of issues and decide common policy actions overcoming conflicting positions. More specifically, treaty making involves agenda setting, collection of information, a forum for interest articulation, a bargaining process, and development of normative statements and rules.

Since the 1972 UN Conference on the Human Environment held in Stockholm, more than 200 legal instruments have resulted from the negotiation of functionally specific arrangements in such areas as ozone depletion, climate change, transboundary fluxes of airborne pollutants, endangered animals and species, marine pollution and so forth (UNEP, 1997, p. 130). The process of international environmental negotiations can remain open ended with the participation of multiple actors and consensus decision making rule. It took over three decades for more than 150 states to reach a comprehensive agreement in the negotiations of Law of the Sea Treaty. In general, lengthy and complicated negotiations can be related to a broad scope of the agenda, a large number of participating parties, and high stakes of the issues. This chapter examines factors relevant to understanding the process of reaching a multilateral environmental agreement.

THE PROCESS OF NEGOTIATION

During the prenegotiation stage, an issue becomes defined and framed on its way onto the official agenda. In considering time and resources to be invested for multilateral negotiation, the issues need to acquire justification for serious discussion. Moreover, functional scopes of negotiation have to be determined, along with the identification of parties, for bargaining. Interest articulation and aggregation have an impact on institutional bargaining by not only revealing the nature and extent of the problems and their main causes but also suggesting possible actions.

Prenegotiation

Most conferences start out with a provisional agenda to prevent a long drawn-out process to make a decision on the items for conference discussion (Kaufmann, 1988, p. 35). No less importantly, a prenegotiation stage determines the timetable, the composition and credentials of delegations, their voting power, conference rules and structure, and decision making and consultation functions of various mechanisms. In setting up the format for negotiation, the advantages and disadvantages of strategies to seek consensus on comprehensive issues with universal participation can be weighed against those of negotiating independent packages in narrow domains.

Agenda formation covers such activities as the expression of major concerns and the exploration of certain issues before official negotiations. In order to move negotiations forward, it is critical for the agenda to be structured and shaped. In the process of agenda setting, issues are narrowed down for focused discussion. Priorities are laid out in separating important issues from less important ones.

Stakeholders are invited to prenegotiation sessions for presenting their interests and goals.

The assessment of environmental changes and the review of their socio-economic impacts precede the formulation of appropriate policy responses. Ideas need to have a wide appeal in order to draw serious support for being included in negotiation. Joint scientific activities, timely dissemination of information, and forums for direct communication on technical issues are important elements during prenegotiation. For instance, timely dissemination and exchange of technical information on the extent of pollution were critical to the initiation of the negotiation on acid rain. A series of informal workshops enables diplomats to get acquainted with scientific expertise and to be introduced to hypothetical questions and creative answers. The role of these workshops can be extended to advisory function with the formal opening of negotiation.

Prenegotiation assessment often proceeds with the creation of intergovernmental panels and working groups. As part of a long educational process, the information sessions organised by working groups raise the awareness about the problems and examine the causal mechanisms relevant to the formulation of action strategies. Studies commissioned for data gathering and analysis affect bargaining dynamics by providing background information for negotiation agendas. Once ideas assume the form of negotiating agenda items, participants formulate their positions on various aspects of the issues.

Negotiation Activities

Bargaining activities accompany issue definitions and the establishment of technical rules and the structures of multilateral decision making. A formal conference structure can be divided into plenary meetings, committees, subcommittees, working parties, and drafting groups. The need for more focused discussion with the disaggregation of complex problems at a small working group level becomes apparent soon after the debate of agenda items at a general conference session. Fulfilling the official deadline of various conference functions largely depends on how well various committee and group activities are organised (Kaufmann, 1988, p. 36).

Various groups are formed to co-ordinate diverse views and prepare proposals that are put together in the package reflecting different interests. Barriers to communication for the structured intergovernmental negotiations are compensated for by informal contact (Benedick, 1993, p. 238). Mutual cooperation among politically divided parties may be sought through interaction at a low profile forum which discusses technical matters. Interactions among various negotiating blocs which defend similar political interests can be made through co-ordinators mandated to represent their views.

Many conference sessions are held behind closed doors given that an open forum can hinder discussion of options for compromise. In accepting or rejecting proposals, however, various domestic interests need to be aggregated. A political process of negotiation involves coalition building in support of competing proposals. Creative solutions are sought through trade-offs between different options. Resistance to costly measures can be reduced with shared sacrifice and a promise of joint gains.

During the course of formal bargaining sessions, the preliminary negotiating drafts prepared before the arrival of delegates often serve as a basis for formal discussion. Prior to a plenary session, it can be revised and elaborated by the chairperson of the conference in cooperation with the conference secretariat. In a response to new issues emerging at a bargaining process, a sequence of texts is produced. While the agreement on some provisions is easily confirmed, the contested provisions draw most attention. Problem-solving sessions would help clarify conflicting interests and expose the range of choices available to the international community. It is unlikely to see quick negotiations with decisive results in the event of the range of vital interests at stake (Sebenius, 1993, p. 191).

The core of the final negotiation stage lies in putting many parts of agreements together to find a comfortable formula. The creation of a single text leads to the effective debates by avoiding separate discussion about mutually exclusive, multiple drafts (Benedick, 1993, p. 238). Multiple proposals by independent negotiating groups have to converge and be integrated by a high level conference body that is not associated with any protagonist. In such a comprehensive endeavour as the Third UN Conference on the Law of the Sea (1973-82), all the drafts of subgroups were pulled together to create a single treaty text. In this process, certain portions of the text continue to be negotiated to incorporate as many different views as possible. A package needs to be constructed in the way to expand the agreement on key interests supported by major bargaining coalitions to other areas.

Rule Making

The voting rules, adopted in bringing a particular treaty, declaration and resolution into existence, can either be based on consensus or a majority vote. Especially when enforcement mechanisms are weak, an agreement has to be attractive to as many participants as possible. In order to have unanimity among participating states, the agreement may have to be reached by the lowest common denominator with an offer of concessions to dissenting parties. The disagreeable components of a large package can be approved with avoiding public and formal voting on the substantive matters.

In considering difficulties to achieve complete agreements on every issue in multilateral negotiations, consensus thus implies the absence of explicit disagreement. In the event of paralysis caused by veto power, two-thirds of the participating states or other qualified majority voting may prevent the agreement from being held hostage to a few reluctant parties. Consensus rule can also be modified with a provision to opt out of specific clauses while limiting veto power.

In order to induce changes in behaviour, the principles and standards of the agreed formula need to be translated into concrete action plans with a focus on the instruments, time tables and targets for achieving treaty objectives. The parties have to agree with the methods for compliance monitoring, criteria performance and financial vehicles. The details of the agreement are normally drafted in the way feedback is ensured for further negotiation. Through a feedback process, the periodic assessment of progress can lead to preparing a new treaty designed to strengthen or amend the existing policy. Changes in rules and the establishment of new operational procedures can be discussed in light of technological advances and new scientific evidence at the regularly scheduled periodic review sessions set forth by the arrangements in the treaty.

In general, a negotiating process does not necessarily unfold in neatly divided phases, as one phase may merge into another. Various bargaining sessions constitute a long cycle of negotiation on certain key issues. International negotiation can be hampered by weak and unco-ordinated prenegotiation assessment activities, uncertainty of a bargaining situation, a lack of consent on expert knowledge, and inadequate institutional mechanisms to resolve differences. On the other hand, it is futile and unproductive to prepare excessively detailed conference planning for multilateral negotiations, due to their unpredictable dynamics involving diverse agendas and a large number of participants with broad interests.

AGENDA FORMATION

The process of agenda formation is characterised by interest articulation and issue representation. An institutional context of prenegotiation discussion is provided by international organisations. They can be a focal point for environmental action by educating national officials and building a wider transnational environmental constituency. International organisations have never remained purely apolitical and technical, directly contributing to the promotion of certain issues as negotiation agendas with their normative functions in rule making. The ozone depletion issue has remained an important concern due to the efforts made by the

United Nations Environment Programme (UNEP). UNEP's Governing Council called upon governments to reduce the use of chlorofluorocarbons (CFCs).

Negotiations for major international conventions have been sponsored by international agencies. UNEP was mandated to serve as a co-ordinator in overseeing pretreaty scientific activity as well as holding expert meetings and workshops for the ozone negotiation. The climate change negotiation was initially prepared by the Intergovernmental Panel on Climate Change (IPCC) resulting from the joint establishment of UNEP and the World Meteorological Organisation (WMO), while the major administrative functions shifted to the Intergovernmental Negotiating Committee during the official bargaining sessions (Dasgupta, 1994, p. 130). As the activities of UNEP's Ozone Secretariat suggest, international agencies act as an information clearing house by distributing documents from past negotiations and reports on scientific, economic and political developments as well as providing details of each agenda item for upcoming meetings (Downie, 1995, p. 177). With their ability not only to conceptualise and frame the issues but also to set up priorities, they influence defining the scope and nature of the negotiation beyond performing technical roles.

In preparing for a formal negotiation, a coordinating committee, made up of representatives of intergovernmental and nongovernmental organisations, scientific institutions, and national governments, can be established. An expert panel is set up to pursue scientific activities needed for collecting, assessing and synthesising technical material for policy makers and the public. The information about the issues can also be gathered by and exchanged among various parties. An ad hoc working group of legal and technical experts lays foundations for a convention by framing problems and defining issues with the utilisation of collected data. In linking scientific matters to policy formulation, working groups facilitate communication between a negotiating body and a community of experts. Scientific uncertainty and disagreement may be discussed at a common forum with the input of working groups. If necessary, a joint fact-finding effort is made to reconcile contrasting views.

Technical assessment, supported by funding from international organisations or interested governments, was made in advance of the conferences on ozone, climate change, and acid rain. Prior to the beginning of negotiations on a climate change treaty signed at the Rio Environmental Summit in June 1992, the UN Intergovernmental Panel on Climate Change gathered information, co-ordinated research and evaluated preliminary proposals for greenhouse gas reduction (WMO, 1990). Their report, which revealed the trend of the rise in average temperatures and its catastrophic social and ecological impact, created urgency in the negotiation (Feldman, 1995).

New goals can be formulated as a result of working group activities. As the meetings of the Ad Hoc Working Group of Legal and Technical Experts for the Elaboration of a Global Framework Convention for the Protection of the Ozone Layer progressed from 1982 to 1985, the initial discussion about helping governments recognise the threats with cooperative research shifted to a focus on international regulation. The so-called 'Toronto Group' comprised of Finland, Sweden, Norway, Austria, Canada and the United States expanded the discussion by proposing a ban on non-essential uses of CFCs in aerosols. The first serious proposal for international regulation led to the adoption of the 1985 Vienna Convention for the Protection of the Ozone Layer, with more tightened control measures being incorporated in the following agreements.

Scientific meetings contributed to agenda setting by placing 'ozone depletion on the international agenda and attracted representatives of states indifferent or hostile to the issue' (Downie, 1995, p. 175). A special meeting of atmospheric scientists to compare computer models of ozone depletion held prior to the third round of protocol negotiations brought attention to the need for urgent action. Clear scientific consensus broke the political deadlock created by challenges from the opponents of tougher regulations during the previous meetings (Benedick, 1998). A growing consensus on the role of CFCs in ozone depletion made it difficult to oppose regulatory actions. In the ozone treaty negotiations, institutional change stems from altered understandings of the nature of the problem to be solved (Young, 1997, p. 15).

Difficulties in negotiating such areas as desertification, hazardous waste, and tropical timber can be ascribed to the lack of co-ordinated scientific inquiry. For instance, disagreement over the definition of desertification, its causes and extent produced time consuming and frustrating debates in the negotiation of the 1994 Convention to Combat Desertification in Countries Experiencing Serious Drought and/or Desertification, Particularly in Africa (CCD). Throughout the negotiations, divergence in discussion as to the 'global' nature of desertification resulted from the disagreement on a proper terminology. Countries in the South such as Egypt wanted to employ the word 'global' because it emphasises the importance of desertification throughout the world, not just in Africa. The US and UK objected to the term since the word 'global' has a connotation for global warming, implying industrialised countries' responsibility for desert solutions (Long, 1997, p. 94).

Bargaining sessions do not move smoothly if the prenegotiation information sessions fail to establish a common base of understanding among the parties. Weaknesses in the final agreement can be attributed to the failure to define issues properly, narrow the gap between divergent views on technical information, and conceptualise alternative policy options prior to formal negotiation. A lack of jointly co-ordinated assessments results in multiple problem definitions, conflicting policy goals and vague treaty provisions. The work for gathering and interpreting data is critical in weighing merits for different action strategies. Cumulated knowledge from extensive monitoring activities, as were illustrated in the Long-Range Transboundary Air Pollution agreement, helps alter country positions, build consensus regarding the nature of the problem, and recognise the need to address it. Effectively disseminated information generated by joint technical assessment produces political pressure for action as well as enhancing a common understanding of issues at the negotiation table.

Interest Articulation and Issue Representation

Outside experts play a constructive role in educating participants, encouraging changes in their positions, and producing new perspectives. Scientists and environmental nongovernmental organisations have influenced agenda setting by providing expert opinions and raising public awareness. Certain environmental issues are put on the top of the international agenda due to public scrutiny. Perceived seriousness of the crisis, generated by the media coverage of scientific findings, puts pressure on policy makers for prompt discussion about possible remedies. In defining interests, new knowledge can be used to influence the attitudes of political elites and of the public.

Scientists often call for international cooperation by using press conferences and public forums as well as petitions. Scientific information brings to light desperate situations that put hundreds of thousands of species in distinction. The protection of polar bears began to draw attention in response to the sensitivity raised by a small group of Arctic wildlife specialists at a meeting convened by the Conservation of Nature and Natural Resources. In a similar manner, marine biologists expressed concerns about human threats to many species of fish, whales, dolphins, seals, and other sea animals (Hempel, 1996, p. 127).

NGOs utilise various sessions of international conferences to shed light on dominant economic interests and political obstacles. In counterbalancing business influence in global warming, biodiversity and other areas, NGOs attempt to shape a widespread public perception through their educational programmes. In the negotiations that resulted in the 1992 Biodiversity Convention, environmentalists insisted on the measures to preserve natural resources in challenge against biotechnology and pharmaceutical firms that were mainly interested in gaining access to biological and genetic resources. The Convention on International Trade in Endangered Species (CITES) was finally signed in 1973 after the International Union for the Conservation of Nature and Natural Resources called attention to the need for international regulation on the export, transit, and import of endangered animal and plant species and their products.

Whereas NGOs support intergovernmental organisations for new initiatives, they put pressure on their own governments to participate in treaty making activities and also lobby to affect their negotiation positions. Since most governments especially in the West are subject to domestic pressure, the major means to influence their government positions is forging domestic coalitions and mobilising public opinion for certain issues. In addition, NGO representatives may actually attend the sessions of international negotiations as part of a national delegation although their role is limited as an observer.

The Antarctic and Southern Oceans Coalition (ASOC) was critical in the negotiation of the possibility of mineral exploration. Since 1983, their representatives have been sitting on the delegations of Australia, Denmark, New Zealand, France, and the United States in an advisory capacity to discuss future status of the region. In the negotiation of the Environment Protocol of 1991, the NGO coalition's analysis and policy recommendations made an important contribution to defeating a proposal to open up Antarctica to future economic exploitation.

NGOs have long campaigned for an agreement on securing the livelihoods of forest-dwelling peoples in collusion with sovereignty claims of producer countries. In the international struggle to save the rainforest and to empower local inhabitants, NGOs forged a close working relationship with indigenous peoples' groups and lobbied the International Tropical Timber Organisation (ITTO) on behalf of the latter. Their concerns were, to a certain extent, reflected in the International Tropical Timber Agreement reached in January 1994 that contains a reference to the need to accord due regard to 'the interests of local communities dependent on forest resources' (Humphreys, 1996, p. 228).

INSTITUTIONAL BARGAINING

Bargaining can be seen in terms of an intricate decision making process involving actors who pursue their own interests. Competing goals have to be compromised so that expectations can converge on focal points. Incentives for being engaged in a complex bargaining process remain strong as long as a range of feasible agreements is more desirable than an outcome of unilateral actions. Reaching agreements requires the efforts to build consensus on the package of provisions that can be accepted by as many participants as possible.

The process to create mutually acceptable formulas can be fostered by creative problem solving, issue linkages, trade-off, mediating skills of the conference leadership, informal communication in a small group setting, and an incremental negotiation strategy. Forging linkages among separate proposals is essential to breaking impasses and realising joint gains. The delegates' work can be introduced in the way to promote mutual understanding while discouraging inflexible position taking. A private forum can be constructed for a small group of delegates to work out a final agreement to be presented at the main conference session. The leadership has to be capable of persuading the adoption of compromised arrangements.

In the areas of a lowest common denominator, the delegates may address relatively narrow, depoliticised topics. Beyond technical complications, negotiations are often stalled owing to political divisiveness and clashing economic interests and values. Negotiators may initially concentrate on a small number of key problems in formulating solutions. Initial preferences tend to be restructured with the emergence of new information and insights. Later the areas of further agreement can be incorporated in the refinement of final negotiating texts.

Problem Solving

The most desirable situation is to integrate the interests of all the parties and enlarge the pie for mutually satisfying positive-sum outcomes. In most cases, because the locus of the welfare is not known at the outset, it is not easy to deploy bargaining tactics aimed at obtaining the best outcome. When a clearly defined payoff space does not exist, it takes time to move from fixed points of discord to a common point of convergence.

In discussion about different options, those who possess strategic positions can have a disproportionate bargaining leverage. In negotiation for the protection of migratory species, territorial jurisdiction over the habitat for those species has a significant impact on deciding measures to protect migrant routes. The overall distribution of the parties' influence in climate change negotiations was related to their varying contributions to global greenhouse gas emissions (Oberthür and Ott, 1999, p. 269). If the issues are not considered equally salient, it is a challenging task to obtain an agreement seen as fair in cost sharing to every participant. On the other hand, integrative bargaining can be more easily achieved when problems are viewed as common. Serious efforts to produce an integrative outcome are made when all parties prefer striking a bargain to no settlement.

Solutions based on the principles of burden sharing can be used to produce jointly acceptable agreements. Parties with a weak implementation capacity can be allowed to assume obligations gradually. Selective incentives or additional benefits may also be introduced as an inducement to accept an agreement that would otherwise be rejected (Tolba and Rummel-Bulska, 1998, p. 18). In Convention on the International Trade in Endangered Species, those who observe conservation standards are guaranteed access to the international market for wildlife products.

Distributional impediments have to be overcome to satisfy competing interests in a fair, balanced manner. When there was a wide gap in positions between the Toronto Group and the European Community in the negotiation on the Montreal Protocol, an ingenious solution was reached by a formula to combine both national consumption and import demands of developing countries in determining a limit on production of CFCs. If the integration of competing options is not feasible, splitting up issues with the design of alternative or parallel courses of action can be considered.

Creative formulas can be invented to resolve seemingly intractable problems by allowing parties to choose one of multiple control forms yielding similar overall effects. The 1997 Kyoto Protocol on climate change was made more acceptable by providing different options for industrialised countries to fulfil their binding emission reduction obligations. Obligatory parties earn credit for their activities and projects that are directly involved in lowering emissions in other countries. Due to the joint implementation mechanism, the agreement has become more appealing to the parties that face strong domestic opposition.

Issue Linkage

Issue linkage is necessary for and may even be inevitable to bargaining simultaneously on multiple agenda items that have substantive interrelationships. Ideally it is preferable for each issue to be discussed on its own substantive merits, but potential linkages among issues enable negotiators to break the impasses (Sabenius, 1993, p. 200). Expanding the zones of agreement is made possible by trade-offs of constituting elements of bargaining valued differently by the parties. Adding issues means that parties have more to offer each other with new value creation.

In complex negotiations, seemingly unrelated issues can be tied together to invite compound agreements supported by the different coalitions. Shared interests and concerns in key issues need to be recognised in realising large-scale 'horse trading' in a package deal. A comprehensive package of responsibilities and rights of the parties, spelled out in a single convention text of the Law of the Seas negotiation, was crafted by resolving differences in more than two dozen issues in the way the loss in one realm was offset by the gains in others. As preferences of the parties may not be known clearly and can change along with their learning curve, the utility of potential items for issue linkage fluctuates. Their discount rates and time-related payoff values should not be considered static throughout the negotiation.

In the negotiation on the Biodiversity Convention presented at the 1992 Rio Environmental Summit, the main interests of developed countries were to strengthen obligations for biodiversity conservation while obtaining access to biological and genetic resources for biotechnology companies. Developing countries, especially those without financial aid and other compensations, saw these broad obligations unfavourably. Thus, major attention had to be paid to linking access to biological resources with technology transfer, profit sharing and financial assistance (Moremen, 1995, p. 14).

In general, the role of issue linkages is considered to be constructive in making a compromise. In negotiating issues like control over global warming, however, discussion about their causes and solutions was politicised through issue linkages. Many industrialised countries focused on tropical rainforest destruction and population growth in the Third World. On the other hand, developing countries pointed at rich life styles in the North. Deadlocks in some environmental negotiations derive from the fact that Third World countries seize the opportunity to demand more grants, favourable foreign investments, improvement in trade terms, and power sharing in international bodies.

Issue linkage enables reluctant parties to join the table given that the bargaining strength of a disadvantageous party can increase with possibilities of changes in value matrix. On the other hand, it can make negotiation more complicated with the enlargement of the agenda (Susskind, 1994, p. 83). The inclusion of a divisive issue in the agenda adds difficulties in agreeing on the issues otherwise considered to be tractable. Some topics need to be delinked or unpackaged if the repackaged portions are not acceptable to veto groups.

A Consensual Knowledge Base

The process of creating rules is affected by specific sets of ideas (Haas, 1992). Knowledge has a profound effect on people's perceptions and influences the way environmental issues are resolved. The convergence of views, supported by shared knowledge, is essential in reaching agreement. Especially in technically complex issue areas where state preferences have not been established, scientific models play a critical role in inventing new options, consequently influencing the negotiated outcome.

Decisions on allocating quotas or regulatory matters often rely on scientific assessment and interpretation of evidence, as proposals have to be justified by knowledge based claims. Various options have different distributional consequences for each set of interests which different groups of countries are concerned about. Data on pollution levels, meteorological conditions and emission rates are critical in decision making on the allocation of responsibility based on each country's contribution to acid deposition (Levy, 1993, p. 80). Technical questions surrounding calculating the impact of various policy options on chlorine concentrations played an important role in the negotiation of the 1990 London Amendments to the Montreal Protocol. Other examples include negotiation for fisheries management regimes that draw upon the authority of the science of marine biology.

Value assumptions, judgement and subjective choices underlying scientific research suggest that science does not always provide neutral, objective answers (Kütting, 2000, p. 25). Bargaining can be hampered by disputes over technical details associated, for instance, with identifying the agent of environmental pollution. Obtaining scientific consensus can also be made difficult by poor data collection and an inadequate monitoring system. In the case of negotiating the 1985 protocol to Geneva Convention on Long-Range Transboundary Air Pollution, it was not easy to agree on the level of reduction in sulphur dioxide emissions owing to the fact that the effects would not be immediately known.

In general, policy makers want to know definitive answers even when the scientific method is not reliable (Molitor, 1999, p. 220). In the areas where it is difficult to gain scientific consensus, opposing theoretical arguments can provoke contentious political debates about future action strategies (Long, 1994, p. 104). During the negotiation of the 1997 Kyoto Protocol to the Global Convention on Climate Change, the opponents of regulations were able to strengthen their positions due to a lack of firm consensus on the question of human activity's impact on the variability of the climate system. Different emission projections served as a political tool for industrial sectors in climate change deliberations.

Whereas the establishment of a commonly accepted body of data and analysis enables negotiating parties to narrow ranges of uncertainty (Benedick, 1998, p. 314), science does not solely determine such issues as the timing and level of regulations, owing to their implications related to the questions of equity and fairness. Given that the utilisation of scientific findings is socially determined, even the agreement on facts does not immediately translate into unyielding political support. As was reflected in the long resistance from the US government to agree on setting up stringent targets and standards at the global warming conferences, a good deal of credible scientific evidence failed to produce political consensus. A lack of enthusiasm is ascribed to the opposition of those whose interests are deeply entrenched in the existing economic and social practice.

The reduction in the emission of carbon dioxide is unpopular since control over the use of fossil fuels affects industrial production and pleasure of modern life. Any future agreement on a tighter control over global warming faces stiff resistance from business sectors that have vested interests in maintaining or even increasing the consumption level of greenhouse gases. Given political difficulties, consensual knowledge and intersubjective understanding are more easily applied in a narrow problem-solving context, and some observers suggest that scientific knowledge plays a more important role in agenda setting than bargaining sessions.

Conference Leadership

In preparing draft texts acceptable to the majority of the participants, conference chairs and other officials facilitate debate and manage the agendas. Conference secretariats contribute to more effective discussion with their studies, reports and surveys that provide a coherent analytical framework for bargaining on issues at hand. Formal power of key conference officers is normally limited to moderating meetings and ruling on order or procedural motions. In an endeavour to accomplish an objective shared by the majority of delegations, however, they can assist opposing sides in the formulation of compromised proposals. In order to broker different interests, they support bargaining strategies that create a crosscutting coalition.

The performance of conference secretariat head and other leading officials can delay or accelerate negotiation. Presiding officers advance the process by preventing the conference from being derailed by intransigent positions and confrontations. Chairs of the committees divided on subject matters at the Law of the Sea Conference utilised their authority in developing draft texts that represented a step toward consensus. They facilitated integrative bargaining and consensus building on various matters of substance with such techniques as merging and matching positions.

Conference chairs or heads of international organisations that sponsor the conferences can provide leadership in breaking through deadlock in negotiations. They take the lead towards realising the goals aimed at by the framework of an intergovernmental conference. The roles of UNEP Executive Director Mostafa Tolba and the Chair of the Montreal Conference Winifred Lang were critical in hammering out many compromises. The persistence and diplomatic skills of Mostafa Tolba helped the conclusion of the agreements at the series of negotiation sessions on protecting the ozone layer sponsored by UNEP. While urging consensus, Tolba advocated and pressed for a strong treaty by opposing a low-level reduction protocol. At the critical junctures during the Montreal Protocol negotiations, he was a forceful mediator.

Representatives of neutral or smaller states are often chosen to chair negotiations. The examples include Tommy Koh of Singapore for the sea negotiations and Winfried Lang of Austria for the negotiations of the Vienna Convention and the subsequent Montreal Protocol. A partisan posture is incompatible with the position's mediating role and obligation to promote the shared objectives of the delegations. In order to enjoy credibility, a presiding officer has to give up a national representation and should not take positions on substantive matters. Other qualifications include the gift in merging the result of separate proceedings into one single outcome (Lang, 1989, p. 38).

A Small Group Process

The complexities of negotiating in a large group can be significantly reduced by small group sessions which are designed to break the deadlock over issues. This process permits parties to exchange their views freely without committing themselves to formal bargaining positions in a conference meeting. By not being restricted by official settings, the participants feel free to explore possible compromises and use the discussions as the basis for new proposals. A series of group meetings can produce an unofficial draft to be studied at a discussion among delegation heads.

Difficulties in communication resulting from the mere multiplication in the number of recipients of messages can be overcome at small group meetings aimed at building consensus (Kaufmann, 1988, p. 173). A more collegial atmosphere exists at small group working group meetings which maintain some form of informality. Personal rapport, unofficial explanatory memoranda and other subtle forms of communication are contrasted with rigid communication of formal diplomacy. Committee and negotiating group sessions, joined by the heads of selected delegations, were effective in formulating key details of the treaties during the Montreal and London negotiations on ozone depletion. Toward the final stage of meetings to conclude the Kyoto Protocol, several negotiating committees and numerous informal groups discussed many unresolved issues on institutional questions, emission trading, voluntary commitments of developing countries and other complex matters (Oberthür and Ott, 1999, p. 80).

The negotiation on the Antarctic Environmental Protocol in 1991 also benefited from the structure of Special Consultative Meetings (SCM) created following the plenary session. The working draft of the protocol identifying specific priorities and guidelines for the activities resulted from collaborative efforts of the SCM that consisted of working group sessions and simultaneous heads of delegation meetings. Small group meetings facilitate bargaining solutions, but too many meetings can be a source of exhaustion for the delegates, subsequently reducing the quality of conference activities.

Bargaining Blocs

Negotiation practice changes with the shift in the number and composition of groups that attend the meetings. In contrast with bilateral bargaining, the representation of diverse views and the common understanding of the same terms by all the participants are not easy in multiparty bargaining. As the conference size and complexity of the issues increase, it is difficult to engage all the groups in interactive decision making. This situation leads to the regular consultations of country groupings and bargaining between different coalitions.

The free caucusing among like-minded states helps explore common positions and proposals. By promoting interaction among members which share similar interests in a particular issue, bargaining blocs reduce transaction costs at a negotiation that involves a sizable number of parties. The exchange of views becomes easy with the formation of coalitions representing different rallying interests. In dealing with complex issues, coalition structures should be flexible enough to satisfy the concerns of major parties. Inflexibility incurring from too much cohesion hinders the efforts to bring competing interests into a mutually acceptable framework of agreement by making crosscutting coalitions difficult (Hampson, 1995, p. 41).

Country groupings can be formed on the basis of their special interests related to varying political and economic conditions. Coalition groups have been created in global climate negotiations either to support or oppose stringent regulations. The European Union (EU) has been the major industrialised leader in favour of a strong international climate policy. It is, in part, due to their vested interest in reducing fossil fuel consumption, and, in part, due to prioritisation of the issue by environmental groups such as the Climate Network Europe. In order to maintain economic growth, on the other hand, the US. China, India, Japan and Russia accounting for more than half the world's carbon dioxide emissions from fossil fuels, have often acted as a blocking coalition against any agreement requiring significant steps toward the mitigation of global warming at various international conferences.

As their common concern outweighs other interests, multilateral and bilateral contact groups can move across established political relationships in exploring common ground. In the Kyoto Protocol process, an informal coalition of the United States, Japan, Canada, Switzerland, Norway, Australia and New Zealand held a general opposition to stringent commitments for greenhouse gas emission reduction favoured by the EU. The Alliance of Small Islands States (AOSIS) which feel threats from a rise in the sea level took initiatives and joined forces in calling for major reductions in industrialised nations' emissions of greenhouse gases. They were supported by other developing countries (excluding oil-producing countries) mostly concerned about equity. However, China and India, whose emission rates are already high and are even increasing, provided stiff opposition to any kind of methods for major cuts which will impact them.

Bargaining blocs articulate the major concerns at stake and ensure their voice being seriously taken into account in the final agreement. Asymmetric bargaining positions derived from structural weaknesses in overall power relations can be offset by coalitions. The formation of coalitions among developing countries helps equalise their power positions at the bargaining table. In the negotiation of the International Tropical Timber Agreement, Malaysia and Indonesia, as major exporters in the timber trade, formed a major voting bloc. During the negotiation, they took a preemptive action by initiating exporting countries' proposal in fending off developed countries' criticisms.

Although coalitions of various kinds can help overcome disadvantages in the size and other dimensions of multilateral negotiations, even a small blocking group can have substantial sway over a range of issues especially under a consensus decision making rule. Like we have seen in the positions of Japan and Norway during whaling negotiations, the insistence on the protection of minor interests of a small group hinders a meaningful agreement.

Sequencing and Incremental Strategies

It is a very time consuming, complicated task to develop a single negotiating text with an agreement on rules that require dramatic changes in human behaviour and state policies. In particular, consensus on package deals is difficult to attain in a universally inclusive process as to issues and participants. A general, non-binding commitment is inevitable with a lack of political consensus needed for drafting specific, detailed obligations. The agreement is easier to obtain by bracketing off the hard decision about precise action strategies. In a two step approach, sequencing tactics are applied to the areas that need increasingly stringent and comprehensive controls but face stiff opposition.

The convention-protocol negotiating process starts with the formulation of conventions that can be regarded as a starting point for developing more specific agreements. The general, non-binding obligations contained in framework conventions are negotiated in the anticipation of a future agreement on detailed implementation strategies with specific targets and timetables. Vague and open-ended agreements allow bargaining flexibility, while leaving room for continuing negotiations. New measures are discussed at regular conference sessions of the parties to monitor and assess the evolving situations. The provisions can be tightened in face of fresh evidence and a shift in public opinions.

Partial or limited settlement is easily achieved in the areas where targets and standards can be satisfied at a lower cost. Many of the drafting exercises for controlling marine pollution, ozone depletion, acid rain, and climate changes were deliberately designed to incorporate a process by which negotiations begin with easier subjects and move on to more difficult ones. The adoption of a loose umbrella agreement controlling acid rain in 1984 progressed to a protocol setting up a multinational monitoring network, then cuts in emissions of sulphur dioxide in 1985, and nitrogen oxides in 1988.

The structure of the agreement would not be complete without establishing more concrete goals and binding obligations. Stringent measures have to wait for neutralisation of opposing forces and solidification of scientific consensus. The specification of overall issues in question or identification of narrow sub-issues is spelled out in protocols. The amendments and annexes would require signatories to take further actions to ensure compliance. It may take several years to negotiate specific terms in protocols moving beyond framework conventions.

Since all the aspects of a complex problem cannot be easily solved in one framework, an incremental approach has advantages with preventing opposing interests from blocking the agreement. Especially when a comprehensive accord is unattainable, joint efforts for the later negotiation of more specific protocols can be set in motion with the legitimisation of the concerns. A modest, early treaty with provisions for review procedures is often preferred to facing uncertain prospect of obtaining more sweeping instruments (Sebenius, 1993, p. 211).

On the other hand, there is a danger of justifying a low compliance standard for the sake of reaching an agreement. The obligation is rendered meaningless unless it specifies the time frame and the methods by which the target must be met. In a step-by-step approach, if parties feel comfortable with the status quo, it is not easy to move forward to more comprehensive solutions. Although it has attractive features, an incremental approach would be inappropriate to the regulation of activities which cause severe and rapid ecological destruction.

Negotiation may proceed with a less than an optimal number of participants in the areas of urgent priority. The early part of the ozone negotiation involved two dozen states with developing countries noticeably missing, but the number of the participants was expanded to more than 60 countries in the negotiations on the protocol and further amendments. The enlargement of the process is inevitable since environmental treaties become ineffective: without universal participation, nonparties undermine the impact of the agreement as free riders (Benedick, 1993, p. 240). Obviously the ozone negotiation structure can be compared with the design of the climate negotiation in which more than one hundred states participated from the beginning.

As it is often pointed out with the use of such examples as the Third Law of the Sea Conference, a strategy to seek comprehensive agreements on agendas with a broad scope runs a risk of managing excessive complexity and delay. In considering that the approach of universal participation and strategies can be a long and drawn-out process, a small scale but expanding agreement may be preferred (Sebenius, 1993, p. 213). A static solution is avoided in favour of designing 'a dynamic and flexible instrument capable of responding to changing circumstances' (Benedick, 1993, p. 243). In lieu of seeking encompassing obligations in a general framework, independent treaties on specific subjects can be pursued. The success in separate agreements on a series of strategically selected areas can be used to build political momentum with further commitments to search for more encompassing solutions. The disadvantages of this approach would be difficulties in lashing different agreements together in an unwieldy negotiating bundle (Sebenius, 1993, p. 198). Separate accords may prove fruitless until they are linked to future negotiations.

CONCLUSION

The successful ending of institutional bargaining proceeding under a consensus rule requires yielding 'contractual formulas acceptable to all the relevant parties' (Young, 1994, p. 133). The outcome relies on a combination of factors, including input from NGOs and the public, the existence of a consensual knowledge base both in scientific and policy-making communities, the role of issue coalitions, and negotiation strategies. A weak final agreement can be attributed to polarised positions among rigid negotiating blocs, a weak political clout of the conference leadership, and insufficient technical input in advancing the agenda. The joint outcome with a positive sum agreement results from an emphasis on shared goals and trade-off between losses and gains. Environmental negotiation entails a political process of decision making on burden sharing as well as normative and cognitive understanding of the issues.

Compared with the negotiation on the ozone layer, an effective agreement on global warming has been more difficult to obtain. It is due to the fact that serious measures to curb the green house effects have broad social and economic repercussions resulting from changes in the levels and patterns of energy use. As the comparison suggests, a bargaining situation can be complicated by the extent to which powerful interest groups are hurt as well as the level of technological advances such as feasibility in the replacement of pollutants.

Environmental issue areas have their unique functional scopes, geographical domains, ideals and material conditions. The positions of parties can alter with their goals, interests, knowledge, technical capacity and power relations in a specific domain. A broad socio-political context of institutional bargaining is constituted by value differences, the availability of scientific information and the range of vital interests at stake.

REFERENCES

R.E. Benedick, 'Ozone Diplomacy', in G. Sjöstedt (ed.), *International Environmental Negotiation* (Newbury Park: Sage, 1993), pp. 219-243.

R.E. Benedick, *Ozone Diplomacy* (Cambridge: Harvard University Press, 1998).

J. Bernstein, et al., 'Summary of the Fourth Session of the INC for the Elaboration of an International Convention to Combat Desertification', *Earth Negotiation Bulletin*, 4 (44) (1994).

C. Dasgupta, 'The Climate Change Negotiations', in I. Mintzer and J.A. Leondard (eds), *Negotiating Climate Change* (Cambridge: Cambridge University Press, 1994) pp. 129-148.

D.L. Downie, 'UNEP and the Montreal Protocol', in R. Bartlett, et al. (eds), *International Organisations and Environmentd Policy* (Westport: Greenwood Press, 1995) pp. 171-186.

D.L. Feldman, 'Iterative Functionalism and Climate Management Organizations', in R.V. Bartlett, et al. (eds), *International Organizations and Environmental Policy* (Westport: Greenwood Press, 1995) pp. 187-208.

P. Haas, 'Introduction: Epistemic Communities and International Policy Coordination', *International Organisation*, vol. 46, no. 1, (1992) pp. 1-36.

F.O. Hampson, *Multilateral Negotiations* (Baltimore: Johns Hopkins University Press, 1995).

L.C. Hempel, *Environmental Governance: The Global Challenge* (Washington, D.C.: Island Press, 1996).

D. Humphreys, 'Hegemonic Ideology and the International Tropical Timber Organization', *The Environment and International Relations* (London: Routledge, 1996) pp. 215-233.

J. Kaufmann, *Conference Diplomacy: An Introductory Analysis* (Dordrecht: Martinus Nijhoff Publishers, 1988)

G. Kütting, 'Distinguishing Between Institutional and Environmental Effectiveness in International Environmental Agreements', *International Journal of Peace Studies*, vol., 5, no. 1 (2000) pp. 15-34.

W. Lang, 'Multilateral Negotiations: The Role of Presiding Officers', in F. Mautner-Markhof (ed.), *Processes of International Negotiations* (Boulder: Westview Press, 1989), pp. 23-42.

M.A. Levy, 'European Acid Rain: the Power of Tote-Board Diplomacy', in P. M. Haas, et al. (eds), *Institution for the Earth* (Cambridge: The MIT Press, 1993), pp. 75-132.

M. Long, 'Expertise and the Convention to Combat Desertification', in L. Susskind, et al. (eds), *Innovations in International Environmental Negotiation* (Cambridge: The Program on Negotiation, 1997), pp. 88-111.

M.R. Molitor, 'The United Nations Climate Change Agreements', in N.J. Vig and R. Axelrod (eds) (Washington, D.C.: *Congressional Quarterly Press*, 1999) pp. 210-235.

P.M. Moremen, 'Negotiating the Biological Diversity Convention', Working Paper, (Program on Negotiation at Harvard Law School, 1995).

S. Oberthür and H. Ott, *The Kyoto Protocol: International Climate Policy for the 21st Century* (Berlin: Springer, 1999).

J. Sebenius, 'The Law of the Sea Conference: Lessons for Negotiations to Control Global Warming', in G. Sjöstedt (ed.), *International Environmental Negotiation* (Newbury Park: Sage, 1993), pp. 189-216.

L.E. Susskind, *Environmental Diplomacy* (New York: Oxford University Press, 1994).

L. Tamiotti and M. Finger, 'Environmental Organisations: New Roles and Functions in Global Politics', *International Journal of Peace Studies,* Vol. 5, no. 2 (2000).

M. Tolba and I. Rummel-Bulska, *Global Environmental Diplomacy* (Cambridge, MIT Press, 1998).

UNEP (United Nations Environment Programme), *Global Environment Outlook* (New York: Oxford University Press, 1997).

WMO (World Meteorological Organization), Intergovernmental Panel on Climate Change: First Assessment Report Overview (Geneva: WMO and UNEP, 1990).

O. Young, *International Governance: Protecting the Environment in a Stateless Society* (Ithaca, NY: Cornell University Press, 1994).

O. Young, 'Rights, Rules and Resources in World Affairs', in O. Young (ed.), *Global Governance* (Cambridge, The MIT Press, 1997) pp. 1-24.

5 Scientific Uncertainty in Environmental Negotiations: The Responses of the Nongovernmental Sector

Pamela S. Chasek

The negotiation of international environmental treaties and agreements takes place within a constantly shifting drama where issues of science, policy and politics arise and interact on a broad landscape. In general, environmental problems are rarely fully understood at the time political decisions must be made. International environmental negotiations often begin before conclusive scientific evidence is at hand. Additional research is required both during the initial negotiations as well as during the implementation phase where there is a need to continually review new scientific findings and implementation data and adjust the treaty accordingly. As a result of the scientific uncertainty that surrounds international environmental negotiations, participating governments often go in search of scientific knowledge and policy advice to help them shape their negotiating positions. Many government delegates often turn to the nongovernmental sector for such information.

The nongovernmental sector has become increasingly involved in international environmental negotiations, especially those that take place within the UN system. Over the past 25 years, the UN has become much more receptive to the participation of a wide range of interest groups who seek to monitor and influence the negotiations. Within the nongovernmental sector three major categories of interest groups provide scientific information to governments, their delegations and the negotiating process as a whole: the business community, the scientific community and the environmental community. Each of these groups has different interests in the issue under negotiations and the outcome that tend to shape the information they provide to governments. Within each of these three categories, disagreements, sometimes serious ones, may arise among individuals and institutional members. On the whole, however, intragroup disagreements are not as influential in the negotiations as the intergroup divergences of interest. The bottom line, however, is that in their attempts to mitigate scientific uncertainty, the nongovernmental sector tends to 'fan the flame' of uncertainty as political agendas inevitably control the scientific ones.

This chapter examines how different nongovernmental actors – specifically the business, environmental and the independent scientific communities – influence the negotiation of international environmental agreements in their role as providers of scientific information. Using case studies of the Montreal Protocol on Substances that Deplete the Ozone Layer (1987) and the United Nations Framework Convention on Climate Change (1992) and its Kyoto Protocol (1997), this chapter argues that by enhancing the level and the quality of the debate surrounding these uncertainties, nongovernmental actors have made a positive contribution to both the intergovernmental negotiating process as well as to the body of science upon which these negotiations are based.

In general, the impact that non-state actors have on the negotiations varies largely based on their technical or scientific knowledge of the issue, their access to decision makers and their ability to bring pressure to bear on their chosen objective. Their level of influence varies from negotiation to negotiation and from topic to topic. Their contributions to the scientific dialogue are often at odds with one another and may actually cause more confusion in the minds and positions of government delegates than not. Yet, when these contributions are added up at the end of the day, these different nongovernmental actors have usually helped to alleviate or mitigate some of the scientific uncertainty surrounding the issue under negotiation, whether it be ozone depletion, climate change, biosafety, marine pollution or biodiversity loss.

SCIENTIFIC UNCERTAINTY

By their very nature, environmental problems do not lend themselves to precision. The causes or long-term effects of a particular environmental problem are not always clear. The non-linear nature of many environmental issues, the large number of common problems and the poorly understood relationship between the natural world and the social world conspire to render global environmental problems difficult to solve (Choucri, 1993, pp. 1-4; Raustiala, 1997, p. 726). As Raustiala (1997, p. 727) points out

> Numerous crosscutting issues exist, and the interactions among trade, development, ecology, and scientific knowledge pose many difficult questions. Most transboundary environmental issues are relatively novel and little experience exists to guide the policy-making process. New problems appear that had never been anticipated or contemplated – such as continental-size holes in the stratospheric ozone layer. This distinguishes environmental cooperation from, for example, cooperation in trade or security, both of which have long histories of international policy making that provide some guidance and benchmarks, if often inadequate.

The degree and type of uncertainty varies from issue to issue. Thomas Homer-Dixon (1993, p. 46-48) outlines four factors that contribute to uncertainty and unpredictability of certain environmental issues. The first is the quality of the theories about the physical and social processes that affect a particular environmental problem. The second is the quality, quantity and even availability of the necessary data. The third is the degree of uncertainty about both the ability and the willingness of humans to change the social, economic, technological, and physical processes that contribute to environmental stress. The fourth and final factor is the degree to which the physical and social processes influencing different environmental issues are chaotic, thereby rendering accurate prediction difficult or even impossible.

Amidst this uncertainty, policy makers are often forced to deal with the issues and determine appropriate solutions. Over the past two decades, more and more environmental problems are being addressed at the international level. This is not only because some of these problems – such as global warming and ozone depletion – are so large that they can be effectively managed only at the international level; it is also because the line between national and international environmental problems is disappearing. Nitrogen oxide emissions, for example, must be regulated locally because of ground-level ozone formation, regionally because of acid rain, and globally because ground-level ozone is a heat-trapping 'greenhouse' gas.

The internationalisation of environmental issues tends to add more uncertainty than it alleviates. In very few cases are all of the scientific issues completely understood and their future implications projected reliably. While scientific measurements often result in the actual identification of the problem that is being negotiated, there are usually more uncertainties than certainties. According to Sjöstedt and Spector (1993, p. 306), 'there are so many scientific parameters of environmental problems that are uncertain and so many correlations that reasonably cannot be stated as causal relationships, that the substance of what is being negotiated, what are appropriate trade-offs, what are reasonable fallback positions, and what are effective outcomes can become rather nebulous.' At the same time, the potential environmental consequences are often catastrophic. Therefore, negotiators are left in the unenviable position of having to negotiate issues that are ill defined. The dilemma they face is negotiating in this state of uncertainty only to find later that the problem was not as significant as once thought or delaying negotiation until substantial scientific results are available only to find out then that irreparable damage has been done (Sjöstedt and Spector, 1993, p. 306).

In spite of this convincing argument about the dilemma negotiators face, not everyone agrees that scientific uncertainty plays a major role in environmental negotiations. Susskind (1994, p. 65) argues that when it comes to bargaining over the actual terms of a treaty, input from scientists is almost always negligible. 'This is because traditional diplomacy encourages the idea that negotiations ought to reflect politics more than anything else.' Susskind goes on to say that once an environmental problem has been defined and the scientists have had their say, 'bargaining tends to be framed mostly in terms of potential economic losses, possible domestic political advantages, and apparent attacks on sovereignty.' Therefore, one can extrapolate from this assertion that negotiations will follow a more or less predictable path without regard for the level of scientific uncertainty or the contributions of the scientific community.

While this argument has a certain degree of merit, it breaks down when one examines some of the evidence. Despite wide initial scientific uncertainty about the problem and its causes, the negotiations on the depletion of the ozone layer did proceed in a forthright way in the late 1970s and early 1980s in an example of a case of 'pre-emptive risk aversion' (Sjöstedt and Spector 1993, p. 306). This case demonstrates that amidst scientific uncertainty delegates can press forward. In the acid rain negotiations between the United States and Canada, on the other hand, the perception of scientific uncertainty was so high on the part of the United States that it results in a veritable impasse in the negotiations. In this case, delegates were apparently willing to suspend judgment and negotiations until the problem and its possible solutions are more fully elucidated by the scientific community. In the ozone case, delegates worked around the scientific uncertainty in the spirit of the precautionary principle. In the acid rain negotiations, delegates used the scientific uncertainty as an excuse for inaction. While politics may have played a major role in these negotiations, scientific uncertainty did have a major impact.

Sjöstedt and Spector (1993, p. 307) recognise the fact that diplomats do not always put scientific evidence or scientific solutions to the fore of their negotiations. Diplomats tend to view environmental negotiations as situations in which conflicting interests must be satisfied through compromise. From their perspective, optimal technical solutions are only as good as the feasibility of concluding multilateral agreements that apply them. Conflicting objectives among governments must be resolved in the formulation of negotiated outcomes. Thus, the role of the negotiator is to cobble together an agreement that maximises national interests, while offering positive payoffs to all the other parties. Negotiation outcomes that fall short of solving the true environmental problems or delay the process, even though they are politically expedient, can be more damaging than no outcome at all. Therefore, one of the biggest challenges in multilateral environmental negotiation is to balance the science and technical aspects of the problem with the process of policy formulation. If this was not of scientific uncertainty that often covers these negotiations like a layer of clouds complicates things even further. Yet it is this uncertainty that often opens the door to the nongovernmental sector and allows their representatives to contribute to the scientific and/or policy debate and, in some cases, to have a direct impact on the formulation of the outcome.

ROLE OF THE SCIENTIFIC COMMUNITY

The scientific community has always had a role to play in intergovernmental environmental negotiations. As the issues under negotiation get more technical and there is greater scientific uncertainty about the possible long-term effects, scientists are relied upon to present the facts and projections, which often set the stage for negotiations. Some argue that it is essential that an international scientific consensus be built that can agree on basic parameters and narrow the ranges of uncertainty to ensure the success of negotiations (Spector, 1992, p. 2). Over the past twenty years, an international network of cooperating scientists and scientific institutions has become a major actor on the negotiations scene. In effect, there now exists a community of scientists from many nations committed to scientific objectivity and welcoming cooperative research, transcending the narrow political and commercial interests of sovereign states. Arguably, the development of high-level computer technology, the Internet and easy international air travel have revolutionised this form of scientific collaboration. Huge teams of scientists can review each other's work, perform integrated assessments and generate ideas that far exceed the aggregation of their particular knowledge. This development profoundly affected both the ozone negotiations and the climate change negotiations.

Traditionally, scientists played a role in the early phases of the negotiating process, yet often faded into the background during the actual bargaining phase. Today this trend has been shifting as changing and improving scientific evidence often emerges throughout the negotiating process and, as a result, has the potential to change the direction of the negotiations. During the climate change negotiations, and in particular the Kyoto Protocol negotiations, and the post-agreement negotiations on the Montreal Protocol on Substances that Deplete the Ozone Layer scientists and scientific findings have continued to influence the nature of the negotiations and the overall understanding of the issue.

The role of scientists in the development of policy to protect the ozone layer offers one of the most dramatic examples of how science can be the driving force behind international policy making. The formation of a commonly accepted body of data and analyses and the narrowing of ranges of uncertainty were prerequisites to a political solution among negotiating parties initially far apart. This community of scientists collaborated with key government officials who also became convinced of the long-term dangers ultimately prevailed over the more parochial and short-run interests of some national politicians and industrialists.

Although the theoretical understanding of the science of ozone depletion progressed considerably since it was first identified in 1974, great uncertainties still remained as diplomats began to negotiate the need for imposing international controls on CFCs in 1986. Thirty years of measurements had not demonstrated any statistically significant loss of total ozone. The seasonal ozone loss over Antarctica was considered an anomaly at the time and could not be explained by the ozone depletion theory. The models could not predict that at existing rates of CFC emissions any global depletion of ozone would occur for at least the next two decades. Nor was their any indication of increased levels of UV-B radiation reaching the Earth's surface (Benedick, 1991, p. 18).

Amidst this uncertainty, diplomats concluded negotiations on the Montreal Protocol in September 1997. Caroline Thomas (1992, p. 225) argues that the compromise was driven by political and industrial imperatives rather than science. The negotiators did not wait for the announcement of the latest NASA scientific reports, which were released just two weeks after the Protocol was signed. Yet, while the role of scientists and their findings might have been repressed in the initial negotiations, there is little doubt that scientists played a much larger role in mitigating the scientific uncertainty surrounding ozone depletion in the post-agreement negotiations phase.

The scientific community and a growing consensus about the CFC-ozone link played a key role in negotiations, making it increasingly difficult for those who opposed action to 'hide behind scientific dissent' (Hampson, 1995, p. 273). According to Eileen Claussen, former US Assistant Secretary of State for Environment, 'science and the consensus among scientists around the world was a critical ingredient in the Protocol process, as was technology, and the consensus that emerged on what could be accomplished and by when' (Tolba and Rummel-Bulska, 1998, p. 84). By 1987 massive ozone loss had been measured by NASA-sponsored high-altitude aircraft examining the stratosphere over Antarctica, and the link between ozone loss and CFCs became impossible to deny. In addition atmospheric and medical scientists were able to draw direct, quantifiable links between ozone loss, increases in surface UV radiation and increased skin cancer. (Paarlberg, 1999, p. 247).

Six months later the fate of CFCs and halons were sealed. On 15 March 1988, the report of the Ozone Trends Panel was released – a 16-month comprehensive scientific exercise involving more than 100 scientists from ten countries, using new methods to analyse and recompute all previous air- and ground-based atmospheric trace gas measurements, including those from the recent Antarctic expedition. The Panel's conclusions made headlines around the world: ozone layer depletion was no longer a theory. Ozone depletion was substantiated by hard scientific evidence and CFCs and halons were implicated beyond a reasonable doubt (Benedick, 1991, p. 110). Thus, the state of the science had fundamentally changed and with it came pressure for a complete phase-out of CFCs and a strengthening of the Montreal Protocol. Further scientific evidence in the subsequent five years led to the complete phase-out of CFCs and halons, along with phase-outs and reductions in the use of a number of other ozone-depleting chemicals as the Montreal Protocol regime was strengthened. Thus, in this post-agreement negotiation phase, it could be argued that consensus within the scientific community – through the mitigation of the scientific uncertainty surrounding the relationship between man-made chemicals and ozone depletion – actually drove the policy makers.

Scientists have also played a major role in the climate change negotiations; however, unlike the ozone negotiations it is the lack of clear scientific consensus that has had a major impact on the process. The Intergovernmental Panel on Climate Change (IPCC), which was established in 1988, has been the principal vehicle for co-ordinating scientific assessments on climate change from around the world. It helped to develop the foundations for identifying both appropriate strategies to address climate change and points of consensus and divergence within the scientific community about the magnitude and scope of climate change. Of the IPCC's three working groups, it was the report of Working Group 1 (assessment of available scientific evidence on climate change) that had the most impact[1]. The executive summary of the report, which was issued in 1990, pointed out that the working group was certain that there is a natural greenhouse effect and that human activities are increasing the atmospheric concentrations of greenhouse gases. Furthermore, the working group calculated with confidence that CO^2 was responsible for over half of the enhanced greenhouse effect in the past and is likely to remain so in the future (Jäger and O'Riordan, 1996, p. 14-15). It was this IPCC report that led to the establishment of the Intergovernmental Negotiating Committee for a Framework Convention on Climate Change in 1990.

The IPCC's influence on setting the international climate agenda was heavily supplemented by informal groups of scientists. For example, prior to the release of the IPCC's first assessment in 1990, more than 700 scientists, including 49 Nobel laureates, petitioned President George Bush to take prompt action on global warming. An even stronger action statement, the 1993 'World Scientists' Warning to Humanity,' was signed by more than 1,600 senior scientists and Nobel laureates (Hempel, 2000, p. 288).

Even once the Framework Convention on Climate Change was signed in Rio de Janeiro in June 1992, scientists continued to debate the extent of climate change and the role that human activities had on the earth's climate. In its Second Assessment Report, published in 1995, the IPCC highlighted a 'discernible human influence' on climate, which went beyond the 1990 report when human influence on climate was something that could not really be concluded at all. The Second Assessment Report stated that the earth's temperature could rise by between 1 and 3.5° C by the year 2010 – an average rate of warming probably higher than any in the last 10,000 years. The report expected temperatures to continue rising after that, even if emissions of the greenhouse gases such as carbon dioxide and methane that trap heat in the atmosphere were stabilised at that time. In spite of its strong assertions, the Second Assessment Report did not represent international scientific consensus.

There are still differing scientific opinions on the causes or even the phenomenon of global warming. Leading scientists, including some of the strongest supporters of the global warming theory, continue to raise questions about scientific uncertainties surrounding climate change. They point out we know very little about the workings of the atmosphere, weather patterns and their effects on the climate. There are also gaps in scientific understanding about thermal absorption by the oceans, poor emissions data, inconsistencies in world temperature measurement, external factors (e.g., dust from volcanic eruptions) and basic design flaws in the climate models themselves (Hempel, 1996, p. 93).

While delegates did adopt the Kyoto Protocol in December 1997, this continued uncertainty resulted in what some environmentalists considered a lack of any sense of urgency in the negotiating process and a weak protocol, where industrialised countries agreed to reduce their collective emissions of greenhouse gases by only an average of 5.2 per cent and many implementation questions were left unresolved. Yet, at the same time, it could be argued that had the Second Assessment Report not highlighted the human influence on climate change, there would be no Kyoto Protocol at all.

In both of these examples, scientists played a major role in setting out the science, albeit with a large degree of uncertainty, that propelled international policy makers into action. Yet, this has proved to be the exception rather than the rule. Although scientists played a supportive and enabling role in the ozone negotiations, on many other issues they remain divided or captured by particular government or private interests, or even their own interests. On some issues, such as biodiversity loss, the whaling ban, the hazardous waste trade, the Antarctic and ocean dumping of radioactive wastes, scientists contributed little to the mitigation of scientific uncertainty because of lack of influence or rejection by policy makers (Porter, Brown and Chasek, 2000, p. 19).

The role of scientists in addressing and mitigating uncertainty during the course of the negotiations can easily be compromised if science and politics are not kept separate and diplomats and other high-level political officials do not act quickly in response to new scientific evidence. The very image of a science separate from the policy process is a feature of the politics of modern science. Professor Bert Bolin (1994, p. 29), a distinguished climate change scientist and former Chairman of the IPCC, noted that the role of science should be 'to delineate a range of future opportunities, and analyse what the implications of development along one course or another might be ... [but] not to recommend one or the other.' When negotiations begin, it is time for delegates to make decisions on this range of future opportunities and that is not always seen as the role of the scientific community. Furthermore, if the scientists are going to be successful in mitigating uncertainty, they have to maintain their political independence so that all parties will respect their results. It is extremely important that they are not seen to be associated with a single policy recommendation or policies of a particular government.

If nations and the general public believe that scientists abuse the trust they place in them – when one so-called expert says yes and an equally distinguished expert says no – science will have no standing in environmental negotiations (Susskind, 1994, p. 71). That is, if anyone can instruct a scientist what to say on his or her behalf and to bend the available scientific methods and evidence to suit his or her political objectives, then the scientific community will be nothing more than another political interest group casting its lot with one coalition or another. Efforts by scientists to mobilise political leaders and the general public, as they have in the climate change debate, marks a critical transition in the role of many scientists from detached observer to passionate advocate (Hempel, 2000, p. 288).

In the marine science community, for example, scientists can be considered a lobby group in their own right. This is because almost all scientists involved in the development of the 'prediction' and 'acceptable damage' policy are based in marine science laboratories; and these laboratories rely heavily upon government funding and contracts to finance ship-time, monitoring and research programmes. Therefore, any shift away from dumping toward either land-based alternatives or waste-prevention technology, means less sites to be monitored and a reduction in status, if not in jobs themselves. According to Stairs and Taylor (1992, p. 122-3), 'it is in our perception that the strongest defenders in the pro-dumping lobby have been not government regulators or industrialists intent on cheap options, but marine scientists with a lifelong record of involvement in dumping programmes. These scientists have used the aura of scientific complexity and "objective" decision making to further their own "interests".'

The second type of difficulty that the scientific community may encounter is that government delegates do not always respond to the latest scientific findings, especially if it may lead to a drastic change in policy. Even though a number of legal instruments, including Principle 15 of the 1992 Rio Declaration on Environment and Development, embrace the 'precautionary principle', governments have not always supported it (Sands, 1999, p. 129). Principle 15 states that lack of full scientific certainty shall not be used as a reason for postponing cost-effective measures to prevent environmental degradation. However, when it comes to presenting new evidence that may mitigate some of the uncertainty surrounding the particular environmental problem, government delegates often turn a blind eye. It takes time for governments to evaluate new scientific findings and to determine what kind of an impact these findings will have on existing governmental policies, the economy and the environment. Rather than reverse existing positions to reflect any new science, governments prefer to take a more cautionary approach, which delays any impact that the scientific community may have on international policy.

ROLE OF THE ENVIRONMENTAL COMMUNITY

The influence of environmental NGOs on international environmental negotiations[2] has been based on one or more of three factors, according to Porter, Brown and Chasek (2000, p. 61). First, the environmental community tends to have expert knowledge and innovative thinking about global environmental issues, acquired from specialising in the issues under negotiation. Second, they tend to be dedicated to goals that transcend narrow national or sectoral interests. Third, environmental organisations, particularly in developed countries, represent substantial constituencies within their own countries that command attention and can sometimes influence tight electoral contests.

In many cases of multilateral environmental negotiation, environmental NGOs have helped to mitigate scientific uncertainty by fielding 'critical' scientists as advocates and, in some cases, by funding independent research. This may take the form of reviewing reports that governments have a vested interest in ignoring or painting in a more positive light than is justified, or it may focus on the uncertainties that are given inadequate representation in models of future impact. In the case of the 1972 London Convention on the Prevention of Marine Pollution by Dumping of Wastes and Other Matter, Greenpeace first fielded expert scientists in the special group set up to examine the scientific aspects of the radioactive waste dumping controversy following the vote for a moratorium. These experts played a key role in highlighting uncertainties in the model of projected impacts, as well as making clear the consequences of past dumping and publicising previously undisclosed dumping programmes. The political consequence of these disclosures was a vote to continue the moratorium (Stairs and Taylor, 1992, p. 123).

Environmental NGOs also address issues of scientific uncertainty by acting as a bridge between those who understand environmental problems in an ecosystemic framework and those who understand it in diplomatic or economic terms (Princen and Finger, 1994, pp. 222-223). In general, those who best understand the biophysical realities – ecologists, indigenous peoples and others who interact closely with natural systems – are not necessarily those inclined to challenge state-sponsored research funding sources or to act politically in the international system. On the one hand, NGOs often conduct their own scientific research in part to counter this tendency toward state dependency within the scientific community. With respect to indigenous peoples and others such as farmers, pastoralists and fishers, argue Princen and Finger (1994, p. 223), the knowledge that supports their livelihoods is grounded in the characteristics of the natural resources. 'But this knowledge is not readily transferred to the dominant political systems. Consequently, many NGOs can operate at the grassroots level or link up with those who do. This helps them ground their activities in such knowledge.' Thus, a critical role of environmental NGOs is to link the essential knowledge base (scientific and earth centred) to the world of politics and to translate 'biophysical needs' into choices a wide range of actors can make at the international, national and local levels.

Within the context of the ozone negotiations, environmental organisations – particularly those in the US – played an important role in providing scientific information and trying to alleviate the stalemate that developed as a result of perceived scientific uncertainty. Environmental organisations such as Friends of the Earth, Greenpeace and the Natural Resources Defense Council educated both the public and policy makers by publishing studies, holding press conferences and funding research. In 1986, the World Resources Institute in Washington hosted a meeting on the ozone issue for European environmental groups. While the negotiations were underway in 1987, representatives of several US NGOs travelled to Europe and Japan to stimulate local environmental groups to take a stand on ozone layer protection to offset the influence of industry (Benedick, 1991, p. 28).

In both the negotiations leading up to Montreal and in the post-agreement negotiating period, environmental organisations worked hard to provide scientific and other information to delegates that countered the position of industry groups. Friends of the Earth, for example, put together presentations and briefing material based largely on the chlorine-loading models employed by the UN Environment Programme. Others focused on reciting examples of misleading industry statements and questionable activities (Benedick, 1991, p.166). These actions often proved critical to the successful outcome of the negotiations. Oppenheimer (1990, p. 345) argues that during the Montreal Protocol discussions the NGOs 'unique viewpoint probably swayed negotiators on not a few points.' NGOs have also played an important role in educating consumers and delegates about the dangers of methyl bromide (the second most widely used insecticide in the world by volume) and hydrochlorofluorocarbons (HCFCs) to the ozone layer.

Nevertheless, there are some who maintain that NGO contributions to the ozone negotiations should not be overstated. Peter Haas (1992, p 218), for example, argues that 'public sentiment and the activities of nongovernmental organisations such as Friends of the Earth had little impact on the adoption of CFC controls. Instead, they tended to merely reinforce government regulations that had already been introduced'.

In the climate change negotiations, environmental NGOs developed a number of independent scientific analyses. Through their continuing access to outside experts, NGOs were able to get immediate second opinions on some of the more complex issues, making it much easier for NGOs to resolve some issues (Rahman and Roncerel, 1994, p. 248). Greenpeace produced one of the early syntheses on the science of climate change, published in the 1990 volume *Global Warming*, edited by Jeremy Leggett. Greenpeace further developed and maintained special relationships with some developing countries and worked with them on understanding the scientific, policy and strategic issues (Rahman and Roncerel, 1994, p. 245).

The Climate Action Network, consisting of NGOs from both developed and developing countries, played an active role in the negotiations by contributing to the policy debate, critiquing drafts emerging from the negotiations and proposing alternatives. They organised seminars, forums and meetings in order to provide an opportunity for a more flexible exchange of points of view. The NGOs also published a daily bulletin, *ECO*, which reported on the work carried out in the various working groups and offered opinion pieces and articles on the latest scientific evidence on global warming.

Another important role played by the NGOs was to keep the most up-to-date information on the science of climate change in the forefront of the negotiations. NGOs brought independent scientists, including James Hansen of NASA, to deliver seminars during the meetings of the INC (Rahman and Roncerel, 1994, p. 265). These seminars raised the profile of the science in the diplomatic process, helped to address and overcome areas of scientific uncertainty, and helped keep attention focused on the objectives of the negotiations rather than just the details of the negotiating process.

In the Kyoto Protocol negotiations, NGOs produced studies that warned that governments in most countries are not likely to meet their emissions stabilisation targets. All NGO evaluations have pointed out the weakness of governmental resolve in alternative planning for high growth sectors such as transport. According to Subak (1996, p. 60), the most 'technically sophisticated environmental organisations have replicated official projections and have also provided their own evidence as to why their country is unlikely to meet a target in view of stated policies and the most recent information on economic growth'.

There is no concrete evidence to say that environmental NGOs have helped to mitigate the scientific uncertainties surrounding climate change. However, NGOs have succeeded in exposing both negotiators and the public to scientific, technical and implementation data. This information may have helped convince many government delegates to the first Conference of the Parties, held in Berlin in March 1995, that negotiations on strengthening the commitments under the Convention were necessary.

It is impossible to state definitively what effect environmental NGOs have had on mitigating scientific uncertainty in environmental negotiations. On the one hand, it can be argued that environmental NGOs have provided a balance to the scientific information espoused by industry representatives. In this case, the results may not necessarily be as pro-active as the NGOs would have liked, but at the same time they may have prevented the negotiations from falling into the hands of industry and those countries who do not see climate change or ozone depletion as serious environmental problems. On the other hand, environmental NGOs have been able to provide government delegates with important scientific and social scientific information that they normally would not have the opportunity to review. Their independence ties to the grassroots and indigenous communities, and their ability to publish reports and newsletters during negotiating sessions increases the value of their input.

There are also a number of government delegates who do not always have access to or understand the most reliable scientific findings. NGOs have the ability of taking this information and presenting it in the form of briefing papers, reports and audio-visual presentations that delegates can read easily during the negotiations. NGO information has often shaped the positions of some of these delegations – particularly the small island states in the context of the climate change negotiations. One of the most prominent examples of this is the interaction between the NGO FIELD and the Alliance of Small Island States (AOSIS) during the climate change negotiations. FIELD has been instrumental in providing AOSIS with advice and legal expertise, greatly enhancing AOSIS's influence in both the negotiation of the Convention and the Kyoto Protocol. In fact, AOSIS tabled the first draft of a protocol, with assistance from FIELD (Raustiala, 1997, p. 728).

By providing extensive information, evaluations and legal opinions, NGO policy research permits governments to redirect scarce resources elsewhere, and provides perspectives and ideas that may not have emerged from a bureaucratic review process (Raustiala, 1997, p. 728). If the medium shapes the message, the impact of NGOs on the negotiations has the potential of becoming a force to be reckoned with.

ROLE OF BUSINESS AND INDUSTRY

Private business and industrial firms, especially multinational corporations, are important actors in international environmental negotiations because of the commercial and economic implications of the outcome. For the most part, they oppose national and international policies that they believe would impose significant new costs on them or otherwise reduce expected profits. Although some business leaders have become advocates for sustainable development, argue Porter, Brown and Chasek (2000, p 71), 'corporations have worked over the years to weaken several global environmental regimes, including ozone protection, climate change, whaling, and international toxic waste trade, and fisheries'. Yet, when they face strong domestic regulations on an activity with a global environmental dimension, corporations are likely to support an international agreement that would impose similar standards on competitors abroad. Furthermore, they may prefer an international agreement if it has weaker regulations on their activities than those that might be imposed domestically.

Nevertheless, the business community cannot be viewed as a monolith in opposition to international environmental agreements. In some cases, companies may see a positive stake in such an agreement. In an example cited by Porter, Brown and Chasek (2000, p. 72), the Industrial Biotechnology Association opposed the Convention on Biological Diversity in 1992 fearing that the provisions on intellectual property rights would legally condone existing violations of those rights. But the issue was not a high priority for most of the industry and two of its leading member corporations, Merck and Genentech, believed that the convention would benefit them by encouraging developing countries to negotiate agreements with companies for access to genetic resources. Similarly, industries interested in promoting alternatives to fossil fuels supported the Climate Change Convention. For example, several energy and oil giants launched major initiatives in the field of solar power. They believed that a strong Kyoto Protocol could be the key to take solar and other forms of renewable energy to the next level (Lynch, 1998, p. 18).

Like environmental NGOs, representatives of corporate interests lobby negotiations on environmental agreements primarily by providing scientific and economic information and analysis to delegations that are most sympathetic to their cause. Industry groups are invariably sophisticated and well organised, particularly in their public relations and lobbying efforts. They usually can devote considerably more resources to these efforts than environmental NGOs. In the case of the climate change negotiations, oil and coal interests were very active in advising the US, Russian and Saudi delegations on how to weaken the regime (Porter, Brown and Chasek, 2000, p. 75). In some cases, such as during the ozone negotiations, some of the 'slick publicity' backfired and industry lost valuable credibility when distortions were exposed by environmental groups and scientists (Benedick, 1993, p. 226).

Many industrialists resist changing traditional ways of doing business. They tend to underestimate the seriousness of an environmental problem and exaggerate the costs of remedial action. But if regulatory policies can provide the market with the appropriate signals, the vast financial, intellectual and technical resources of the corporate sector can be stimulated to undertake the needed research and development of environmentally sound products and technologies (Benedick, 1993, p. 227). Along these lines, Porter, Brown and Chasek (2000, p. 75) point to the ozone-protection issue where 'the US chemical industry delayed movement toward any regime for regulating ozone-depleting CFCs in the early 1980s, in part by simply reducing their own research efforts on substitutes.' However, to give the business community the benefit of the doubt, their actions can also enhance compliance to an environmental regime. In the late 1980s and early 1990s, CFC producers gave impetus to an accelerated timetable for the phaseout of CFCs by unilaterally pledging to phase out their own uses of CFCs ahead of the schedule agreed to by the parties to the Montreal Protocol.

There are examples that show that at least parts of industry recognise that self-interest requires acceptance of some responsibility. The chemical industry contributed to studies that led to the phasing out of CFCs in spray cans in the late 1970s. Industrial research scientists, along with university and government scientists, were active in research that led to an understanding of the Antarctic ozone hole. The halocarbon industry committed to working to safeguard the ozone layer by replacing damaging halocarbons with less damaging gases.

In a rare case of cooperation between industry and environmental organisations, both sectors supported the negotiation of the Montreal Protocol, albeit for different reasons. In the late 1970s and early 1980s, a number of US environmental groups put a great deal of domestic pressure on US corporations, municipalities, state legislatures and the federal government to restrict CFC production. In 1980, the US Environmental Protection Agency proposed a no-growth formula that would restrict CFC production at current levels. While US industries successfully fought the measure, it sent a message to them that sooner or later the United States would impose some form of CFC regulation (Wapner, 1996, p. 131).

The threat of domestic regulation scared US industries. If the US was the only country to impose restrictions, then US industries would be at a competitive disadvantage to foreign CFC producers. To put it in more theoretical terms, the interdependencies of the global marketplace threatened to hurt US firms disproportionately. In the face of such circumstances, industrial lobbying groups, such as the Chemical Manufacturers Association and the Alliance for Responsible CFC Policy, began to call for international standards. Arguing that the US economy would suffer disproportionately, these industrial lobbying groups claimed that the US could only call for cutbacks in production if it did so internationally (Wapner, 1996, p. 131). However, at the same time, Dupont, the largest US manufacturer of CFCs had been intensively searching for CFC substitutes, and appeared to be well ahead of its global competitors in this regard. Thus if international limits were placed on CFCs, Dupont would be in a favourable competitive position vis-à-vis the safer alternatives it had been developing (Sebenius, 1994, p. 291; Paarlberg, 1999, p. 246). Thus, paradoxically, US industry was a driving force behind the Montreal Protocol (Wapner, 1996, p. 132).

In the climate change negotiations representatives of the corporate sector play a very different role. After the release of the IPCC's first report in 1990, a number of scientific uncertainties remained. This fact, coupled, with the absence of any unqualified proof that greenhouse-induced global warming was happening, left the door open for politicians and representatives of oil, gas and electric companies who banded together to form the Global Climate Coalition to cite conflicting scientific advice as a reason for a cautious political programme. The Global Climate Coalition argues that climate policy decisions must be made with the benefit of an adequate scientific understanding of the how and why of climate changes. They stress the fact that scientists remain divided on a number of climate change issues, including the role of man-made gases, the causes of global temperature change over the past century, the accuracy of forecasts based on computer modelling, sea-level rise, and the effect of increases in carbon dioxide on the world's plant life. They consistently call for more scientific research to address these uncertainties and that this research should be done before any drastic policy decisions are taken.

Essentially, powerful industrial interests perceived significant costs associated with reductions in carbon dioxide emissions. Although the ozone layer experience might indicate that any initial opposition would eventually give way to industry cooperation, Rowlands (1995, p. 137) notes that significant differences between the two issues challenge this. He continues:

> Recall that, on the ozone layer issue, the large chemical companies eventually supported CFC regulation, because their representatives realised that they would be the ones that would manufacture the substitute chemicals. The collapse of one market, therefore, would be compensated by the emergence of another. In contrast, substitution would not occur to the same extent on the global warming issue.

> Notwithstanding the alternative use of lower-carbon or non-carbon fuels, the primary prescriptions to combat global warming did not involve replacement (as in the CFC case), but rather reduction. Energy

producers therefore did not anticipate 'different' business, but rather 'dissipating' business.

During the final stages of the negotiation of the Kyoto Protocol in 1997, a number of powerful industry groups in the United States mounted a US$13 million advertising campaign to persuade the American public (and, therefore, Congress and the Clinton administration) that efforts to force industry to lower greenhouse gas emissions will cripple the economy. The campaign raised questions about the integrity of scientific forecasts regarding the environmental costs of global warming and claimed that abatement efforts would cost the American economy up to US $275 billion and as many as two million jobs (Lynch, 1998, p. 17). The Center for Energy and Economic Development, a group with a $4 million annual budget sponsored by the coal industry, targeted US business and civic groups with a similar message (Paarlberg, 2000, p. 246). In October 1997, Lee R. Raymond, Chairman and Chief Executive Officer of the Exxon Corporation, speaking at the World Petroleum Conference in Beijing, called upon developing Asian nations to increase fossil-fuel use and to work with Exxon to oppose any international agreement to control greenhouse gases (Lynch, 1998, p. 17).

Over the last decade, industry coalitions have become much more active in international environmental negotiations. Their participation derives from concern that the policy guidelines set out by the international community will have a direct impact on their economic well-being. The energy industry has aligned with the oil producing nations in the climate change negotiations, the chemical industry has actively lobbied delegations during the ozone negotiations and pulp, paper and logging industries have participated in the United Nations' Intergovernmental Forum on Forests, as well as in the work of the International Tropical Timber Organization. The close relationship between economic and environmental concerns ensures that business and industry groups will continue to do their best to influence the negotiation of agreements so that their interests are adequately reflected.

CONCLUSIONS

While the negotiation of environmental treaties and the formation of international environmental policy remains the bastion of governments, a number of different types of nongovernmental actors have been slowly increasing their visibility and activities within and outside the negotiating chambers. This increased role has been the result of a number of events that have come to the fore during the past decade, including encouragement of the participation of the nongovernmental sector by the United Nations system, the increased complexity of the environmental issues under negotiation and the continuing scientific uncertainty surrounding these issues. Government delegates have learned to deal with the increased presence of non-state actors by respecting their opinions, reading their materials, listening to their speeches or ignoring them altogether. Yet, in many cases delegates have found their presence to be useful, especially in terms of the provision of scientific, political, economic, environmental and social information. In the negotiation of environmental agreements delegates often face the dilemma of negotiating in a state of scientific uncertainty. They can either successfully negotiate an agreement only to find later that the problem was not as significant as once thought or they may delay negotiation until substantial scientific results are available only to find out then that irreparable damage has been done. Nongovernmental actors can play an important role in informing the negotiating process by helping to reduce or mitigate this scientific uncertainty.

While the scientific, environmental and business communities attempt to reduce this uncertainty from different perspectives, the information that they provide to government delegates and other participants in multilateral environmental negotiations serves to enhance the level of the debate, increase the level of expertise of the negotiators and expand the overall body of knowledge on the environmental issue under negotiation. The bottom line, however, is that in their attempts to mitigate scientific uncertainty, the nongovernmental sector, through the provision of often contradictory material, tends to 'fan the flame' of uncertainty as political agendas control the scientific ones.

The role of the 'independent' scientific community usually precedes the actual negotiation of a treaty and then usually comes into play during the implementation phase. Often it is scientific evidence that triggers the negotiations in the first place. They do the initial research and prepare the reports that propel governments into action. While not all scientists are as independent as they seem (many are funded by governments, environmental groups and industry), they usually do not actively lobby governments during negotiations. They tend to leave this task to environmental NGOs and industry representatives. Yet, their ability or inability to reach consensus often has a greater impact on the negotiations than any direct lobbying. The absence of scientific consensus, as has been demonstrated by the early negotiations to protect the ozone layer and the climate change negotiations, decreases the ability of delegates to reach their own consensus on the policy measures needed to address the problem and increases the ability of other nongovernmental actors to influence the negotiations.

Both environmental NGOs and industry associations focus more on lobbying government delegations and trying to influence the outcome of the negotiations. Yet, at the same time some of these nongovernmental actors also fund their own scientific research and prepare reports that interpret or critique the findings of other scientists. All of their activities are aimed at convincing delegates that either there is no real scientific uncertainty, that governments should act even in the face of scientific uncertainty, or that the uncertainty is so great that governments may regret any hasty decisions. In the case of the environmental NGOs, their actions are more in line with the mitigation of scientific uncertainty – particularly if it advances their cause. While their research and their publications may not always provide clear-cut scientific evidence, they are usually successful in presenting this material in a manner by which even the least scientifically-educated diplomats can understand the issues at stake.

The business and industrial community, on the other hand, often tends toward the exacerbation of scientific uncertainty since governments are more likely to act cautiously if the uncertainty is greater. The aim of the business community is to prevent governments from adopting an agreement that would impose significant new costs on them or otherwise reduce expected profits. However, as in the case of the Montreal Protocol and among some companies or sectors in the case of the Kyoto Protocol (i.e., British Petroleum, the insurance industry), there have been occasions where an industry may determine that promoting an international agreement will be more beneficial than domestic legislation that might hurt their competitiveness on foreign markets.

These two often divergent interpretations of scientific evidence often creates more uncertainty than they mitigate, especially in the eyes of government delegates who are trying to process this information. There will always be self-interested actors willing to exploit scientific uncertainty for their own ends, arguing against any global action (that would hurt them) on the grounds that a fuller understanding is required before a clear course of action can be charted. When scientists acknowledge uncertainty, they allow political actors great control over decision making (Susskind, 1994, pp. 63-64). As a result, the information provided to delegates is often contradictory and can serve to inflame the debate rather than resolve anything. Nevertheless, this information has the potential of shaping the outcomes of international environmental negotiations and shaping the international community's response to environmental problems for years to come.

NOTES

1. Working Group 2 was asked to assess environmental and socio-economic impacts of climate change. Working Group 3 was to formulate response strategies.

2. The environmental community consists of a wide range of organisations that operate at different levels: international organisations that contain branches in a number of different countries; national or local organisations focused primarily on domestic environmental issues; and think tanks or research institutes whose influence comes primarily from publishing studies.

REFERENCES

R. E. Benedick, 'Perspectives of a Negotiation Practitioner', in G. Sjöstedt (ed.), *International Environmental Negotiation* (Newbury Park, CA: Sage, 1993).

R. E. Benedick, *Ozone Diplomacy* (Cambridge: Harvard University Press, 1991).

B. Bolin, 'Science and Policy making', *Ambio*, vol. 23, no. 1 (1994).

N. Choucri (ed.), *Global Accord* (Cambridge, Mass.: MIT Press, 1993).

R. G. Fleagle, *Global Environmental Change: Interactions of Science, Policy and Politics in the United States* (Westport, Conn.: Praeger, 1994).

P. Haas, 'Banning Chlorofluorcarbons: Epistemic Community Efforts to Protect Stratospheric Ozone', *International Organization*, vol. 46, no. 1 (Winter 1992).

F.O. Hampson, *Multilateral Negotiations* (Baltimore: Johns Hopkins University Press, 1995).

L.C. Hempel, 'Climate Policy on the Installment Plan', in N.J. Vig and M.E. Kraft (eds.), *Environmental Policy*, 4th ed (Washington, DC: Congressional Quarterly Press, 2000).

L.C. Hempel, *Environmental Governance: The Global Challenge* (Washington, DC: Island Press, 1996).

T. F. Homer-Dixon, 'Physical Dimensions of Global Change', in N. Choucri (ed.), *Global Accord* (Cambridge, Mass.: MIT Press, 1993).

J. Jäger and T. O'Riordan, 'The History of Climate Change Science and Politics', in T. O'Riordan and J. Jäger (eds.), *Politics of Climate Change: A European Perspective* (London: Routledge, 1996).

C. Lynch, 'Stormy Weather', *Amicus Journal* (Winter 1998) pp. 15-19.

M. Oppenheimer, 'Responding to Climate Change: The Crucial Role of the NGOs', in H-J Karpe, D. Otten and S.C. Trinidade (eds), *Climate and Development: Climatic Change and Variability and the Resulting Social, Economic and Technological Implications* (London: Springer-Verlag, 1990).

R. Paarlberg, 'Lapsed Leadership: U.S. International Environmental Policy Since Rio', in N.L. Vig and R.S. Axelrod (eds), *The Global Environment: Institutions, Law and Policy* (Washington, DC: Congressional Quarterly Press, 1999).

G. Porter, J. W. Brown and P. Chasek, *Global Environmental Politics,* 3rd ed (Boulder: Westview, 2000).

T. Princen and M. Finger, *Environmental NGOs in World Politics* (London: Routledge, 1994).

K. Raustiala, 'States, NGOs and International Environmental Institutions', *International Studies Quarterly*, vol. 41 (1997), pp. 719-40.

A. Rahman and A. Roncerel, 'A View from the Ground Up', in I. M. Mintzer and J.A. Leonard (eds), *Negotiating Climate Change* (Cambridge: Cambridge University Press, 1994).

W. Rowlands, *The Politics of Global Atmospheric Change* (Manchester: Manchester University Press, 1995).

P. Sands, 'Environmental Protection in the Twenty-first Century: Sustainable Development and International Law', in N.L. Vig and R.S. Axelrod (eds), *The Global Environment: Institutions, Law and Policy* (Washington, DC: Congressional Quarterly Press, 1999).

J. K. Sebenius, 'Towards a Winning Climate Coalition', in I. M. Mintzer and J.A. Leonard (eds), *Negotiating Climate Change* (Cambridge: Cambridge University Press, 1994).

G. Sjöstedt and B. I. Spector, 'Conclusion', in G. Sjöstedt (ed.), *International Environmental Negotiation* (Newbury Park, CA: Sage, 1993).

B. I. Spector, *International Environmental Negotiation: Insights for Practice*, Exec. Report 21 (Laxenburg, Austria: IIASA, 1992).

K. Stairs and P. Taylor, 'Non-Governmental Organizations and the Legal Protection of the Oceans: A Case Study', in A. Hurrell and B. Kingsbury, *The International Politics of the Environment* (New York: Oxford, 1992).

S. Subak, 'The Science and Politics of National Greenhouse Gas Inventories', in T. O'Riordan and J. Jäger (eds), *Politics of Climate Change* (London: Routledge, 1996).

L. E. Susskind, *Environmental Diplomacy* (New York: Oxford, 1994).

C. Thomas, *The Environment in International Relations* (London: The Royal Institute of International Affairs, 1992).

M. K. Tolba and I. Rummel-Bulska, *Global Environmental Diplomacy* (Cambridge, Mass.: MIT Press, 1998).

P. Wapner, *Environmental Activism and World Civic Politics* (Albany: State University of New York Press, 1996).

Part IV: Institutional Context

6 The Legitimacy of the Global Environment Facility

Rodger A. Payne

At the June 1992 United Nations Conference on Environment and Development (UNCED), the Global Environment Facility (GEF) was named the interim funding mechanism for both the Framework Convention on Climate Change and the Convention on Biological Diversity, the two most important international agreements to emerge from the Rio meetings. GEF was also explicitly tagged as the major funding instrument for Agenda 21, UNCED's forward-looking sustainable development blueprint. Consequently, for the impoverished states of the Global South and for the so-called 'economies in transition' of Eastern Europe, the GEF now serves as the principal source of development assistance for global environmental purposes. Over $2 billion worth of grants have been authorised to date and another $2.75 billion has been pledged to cover the next four years (El-Ashry, 1998).

GEF funds are presently at work in about 120 countries, targeted at hundreds of projects addressing the most prominent global environmental concerns, including not only climate change and loss of biological diversity, but also threats to international waters, depletion of the atmospheric ozone layer, and land degradation. The GEF was also charged with 'mainstreaming' global ecological considerations into development planning within both nation-states and partner organisations like the World Bank and United Nations Development Programme (UNDP). In practice, this means that the GEF, by attaching funds to projects financed by others, has leveraged an additional $5 billion worth of development spending (El-Ashry, 1998). Based on the financial record, the GEF emerged during the 1990s as the most influential global environmental institution.

Realistically, however, the GEF's environmental record cannot yet be evaluated. External reviewers, including both independent researchers commissioned by the agency (Executive Summary, 1998) and private scholars investigating GEF goals and operations (Young and Boehmer-Christiansen, 1997, p. 196), agree that it is simply premature to evaluate most GEF ventures. Many original pilot-era projects have only recently moved from the design to implementation stages. Thus, rather than examining some potentially atypical project successes or failures, this chapter considers important questions related to the GEF's institutional legitimacy. Does the world community view the GEF as the proper agency to address global environmental concerns? Will the GEF have lasting and authoritative power in this issue area?

These questions are important because the GEF has a turbulent history. During its first three years of operation, GEF was 'subjected to an extraordinary amount of international scrutiny and criticism' (Fairman, 1994, p. 39), much of it related to the expansive North-South political divide (Gupta, 1995). Remarkably, only a dozen of the 29 pilot stage members were from the Global South. Like the World Bank and virtually all other multilateral development institutions, governance of the original GEF was dominated by affluent donor states. Critics warned that the GEF would thereby inherit the noxious defects of its World Bank financial trustee: excessive secrecy, arrogance, and unresponsiveness (Rowlands, 1995, pp. 203-4).[1] Thus, the GEF 'lacked legitimacy with governments of poor countries' (Anonymous, 1996, p. 18) and with a plethora of environmental advocacy organisations, viewed by many as the GEF's most articulate and vehement critics (Sharma, 1996, p. 77). For these potential clients and concerned third parties, the pilot stage GEF simply did not reflect the interests or needs of most of the world.[2]

Thus, while UNCED impressively named GEF the primary funding mechanism for the new climate and biodiversity conventions, this status was granted only on an interim basis and was made explicitly conditional on a major structural overhaul. Institutional legitimacy, indeed institutional survival and perhaps even the future of truly global environmentalism, hinged on reconstructing GEF's internal workings to the satisfaction of its critics (Jordan, 1994, p. 32; Mertens, 1994, p. 108). As the GEF's Head of External Affairs explained, restructuring had to include provisions for 'universal participation' as well as 'greater transparency and democracy in governance' (Van Praag, 1994). After a thorough external review and a contentious international negotiation, retooling was finally agreed in 1994. The refurbished GEF Instrument includes new guidelines encouraging participation by non-state actors, governance structures granting unprecedented influence to recipient states, more onerous project approval procedures, and requirements for independent monitoring and evaluation of performance (Bowles, 1996). Because of the implications of these reforms, Susskind (1994, p. 94) calls the expansion and modification of the GEF 'probably the greatest accomplishment' of UNCED.

Clearly then, an evaluation of the GEF's legitimacy is a worthwhile endeavour. In this chapter, I specifically discuss the importance of legitimacy in global politics, consider how claims about international institutional legitimacy or illegitimacy might be assessed, and evaluate the authority of the restructured GEF. While many other overviews of the GEF have considered the importance of organisational reform, they focus almost exclusively on the original revisions, overlook recent changes in practice, and do not ponder the importance of institutional legitimacy to GEF's ongoing success (see Gupta, 1995; Jordan, 1995; Sharma, 1996; and Payne, 1998).

LEGITIMACY IN WORLD POLITICS

Political agents and structures are mutually constituted. For the study of world politics, one important implication of this co-dependence is that global actors, including nation-states and perhaps important nongovernmental organisations (NGOs), create norms or institutions that reflect their intersubjective understandings.[3] Unfortunately, as Checkel (1998) notes, scholars employing the constructivist approach have so far given insufficient attention to theorising how agents constitute and alter structures. The empirical work employing this method instead tends to demonstrate how norms affect actor identities and/or practices. Thus, constructivists need to reverse the causal arrow and explain how social relationships undergird changes in norms or institutions like the reconstructed GEF.

One particularly promising means to understand how structures are constituted and change focuses on their legitimacy. This is because the legitimacy of all political orders, except perhaps for those based on unyielding coercive power, hinges on the mutual acquiescence of the relevant constitutive agents (Barnett, 1997). In other words, legitimacy refers to the valid authority granted by the relevant political community and vested in particular political structures.[4] Accordingly, an international institution like the GEF is legitimate if the constituting states (and perhaps important non-state actors) intersubjectively agree that it was formed to achieve proper goals and operates according to appropriate procedures. Restructuring is practically required if an institution is viewed as substantively or procedurally illegitimate. For example, by the late 1980s, Soviet domination of the Warsaw Pact and its own internal republics came to be viewed as an unacceptable structural arrangement (Koslowski and Kratochwil, 1994) and this perception of illegitimacy set in motion tremendous domestic and global structural change. Impotence, of course, is an unsuitable alternative to restructuring. The Commission on Global Governance (1995, p. 66), an august study group comprised of prominent civil servants from around the world, observed recently that international 'institutions that lack legitimacy are seldom effective over the long run'.

I already noted above that the pilot GEF was viewed almost from its founding as illegitimate by the poor states of the Global South and by numerous globally active NGOs. Later in this chapter, I explain in more detail why these actors viewed the GEF as illegitimate, how the institution was reconstructed to mollify their criticisms, and what the future may hold given ongoing concerns about the agency. Initially, however, it is important to investigate the role persuasive discourse plays in the way actors come to agree about what comprises a legitimate structure or institution. So-called 'liberal constructivism,' for instance, explains how advocates might impute a particularly compelling set of agreed understandings.

Legitimacy and Persuasion

A useful explanation of global structural change must examine how norms or institutions are challenged, or contested, by the constitutive political community. The parsimonious view is that agents, hoping to shape and/or reshape mutually agreed understandings, offer alternative ideas and interpretations about rules of mutual interaction, norms, or the purpose and operation of organisations. If these claims are sufficiently persuasive, other actors will at the extreme agree to reconstitute the social structure(s).[5] More frequently, of course, agents will alter their practices to account for subtle new understandings about challenged structures. The latter course is a recipe for slower, incremental change. Modifications in practices will to some extent affect understandings of structures, which should ultimately affect agent identities, preferences, and practices. The result should be recurrence of the entire cycle and potentially even more change. This recursive view of global political transformation supports the contention of Wendt (1987, p. 359; and 1994, p. 391) and others (Litfin, 1994; Payne, 1996; and Samhat 1997) that there is an 'inherently discursive dimension' to understanding how agents and structures are mutually constituted.

Not all arguments are equally capable of provoking change. Some may be ignored or trumped by more effective claims. Many arguments, potentially even very influential ones, will support the status quo. Consequently, constructivists need to evaluate what makes claims 'persuasive and compelling if they are to understand mutual constitution processes of agents and structures at the international level' (Finnemore, 1996, p. 142). Reus-Smith (1997, p. 564), applying ideas from Habermas, asserts that the most persuasive arguments (or discourses) advanced by agents are those that are based on reasons, resonate with higher order values, and appeal to 'deep-rooted, collectively shared ideas'. Unfortunately, a well-known problem with discursive analysis is that any number of appeals can seem subjectively persuasive to an analyst. Since constructivists need to understand which appeals are intersubjectively persuasive, their own interpretations need to be not only convincing, but also supported by empirical evidence. As Risse-Kappen (1994, p. 187) has argued, 'decision makers are always exposed to several and often contradictory policy concepts'. Thus far, he finds, research has mostly failed 'to specify the conditions under which specific ideas are selected and influence policies while others fall by the wayside'. Moreover, Finnemore (p. 130) challenges constructivists to 'provide substantive arguments about which norms matter'.

What makes some discourses particularly effective at generating broad support? The answer offered here centres on actor claims about structural legitimacy. Arguments about the legitimacy of structures, at least potentially, can induce very fundamental changes. When legitimacy is contested, advocates are essentially arguing that a norm or institution has not been mutually agreed. It lacks socially supported authority. Of course, agents cannot merely assert that structures like the GEF are illegitimate. To 'win' any debate, they must both explain their position and champion ideas that other actors find compelling.

Some constructivists have provided useful guidelines for evaluating arguments about structural legitimacy and world politics. Two are explored here. First, *universal membership* (based on the principle of sovereign equality) is offered as a litmus test for institutional legitimacy. Barnett (1997), for instance, contends that the United Nations is a legitimate institution and can more importantly bestow legitimacy on particular practices because of its near-universal state membership. Upon initial consideration, however, this may not be a particularly helpful insight given that the UN is arguably the only current organisation featuring this characteristic. While this standard may work well in explaining the proper authority of UN actions, it is not at all certain that institutional affiliation with the UN necessarily grants legitimacy to separate organisations. Consider the actions of the Security Council, which are frequently disputed because of dominance there by five permanent veto-wielding states. The problem in those cases, however, is the absence of sovereign equality, which therefore renders the example consistent with Barnett's claim. The practices of non-universal institutions may be open to widespread criticism. In any event, universal membership paired with sovereign equality offers a potentially compelling rationale for institutional legitimacy.

A second plausible and useful constructivist standard for evaluating institutional legitimacy involves *fair procedures*, potentially even democratic processes. As Franck (1990, p. 64) succinctly explains, legitimate institutions must 'function in accordance with ascertainable principles of right process'.[6] In particular, Wendt (1994, p. 393) worries about the potential lack of procedural legitimacy in global institutions assuming comprehensive prerogative over particular issue areas or in a given geographic region. Specifically, these supranational entities are bound to be much less democratic than their individual member-states.[7] This incongruity is a frequent concern in the European Union as many analysts charge that the EU is unaccountable, secretive, and unrepresentative (Gamble, 1993, p. 336; Lodge, 1994, p. 344-5), especially as compared to member-state governments. In short, the EU may suffer from a 'democratic deficit' and specific functions assumed by it or its subsidiaries are thereby potentially illegitimate.

In the current global context, large membership institutions like the restructured GEF probably need not be as democratic as the smaller and unique EU. Around the world, many states seem indifferent or even hostile to democratic principles, while this is not at all true in the EU where democratic government is a requirement for membership. Still, the recent scholarship on 'global governance' and the contentious debate about GEF restructuring demonstrate that more than a modicum of democratic decision making is required for broad-based institutions to exhibit procedural legitimacy. Gordenker and Weiss (1996, p. 221), for instance, argue that 'the agreed and proverbial bottom line for all definitions of global governance ... consists of enhanced transparency, accountability and participation'. The influential Commission on Global Governance (1995, pp. 65-7; and Barnett, 1997) supported this claim by linking international legitimacy to democratic procedures like accountability and participation.[8] As noted above, the biodiversity and climate conventions explicitly required the pilot stage GEF not only to achieve universal participation but also to adopt greater transparency and democracy.

Interestingly, 'participation' in global governance now seems to refer not only to unlimited state membership, but also to significant involvement by NGOs, or other non-state actors. The NGOs involved in numerous global forums, especially those headquartered in affluent states, have lead calls for greater transparency and accountability in international organisations. Although some developing states warily view the demands of Northern NGOs, the calls for democratisation of institutions often reverberate in many national delegations and thereby help create and sustain a convincing coalition. Moreover, with ever growing interdependence and globalisation, expanding transnational networks of peoples and groups (Keck and Sikkink, 1998), and a budding global civil society (Lipschutz, 1992), the integration of NGOs into global decision making processes is increasingly viewed by numerous global actors as a veritable precondition for institutional legitimacy (Spiro, 1994, p. 51; Barnett, 1997, pp. 538-9).[9] Below, it is demonstrated that these arguments have been convincingly made in the context of the restructured GEF.

Liberal Constructivism?

The conjoining of fair procedures with democratic principles can also be defended by examining the persuasiveness of NGO claims for donor-nation domestic audiences. A fair amount of constructivist empirical work supports the contention that 'democracies externalise their internal norms when cooperating with each other' (Risse-Kappen, 1996, p. 368; Sikkink, 1993, p. 437). However, this finding need not mean that democratic states literally push for their values in global forums. Instead, Northern NGO critics of international organisations should find that their arguments calling for greater participation and transparency are especially persuasive in democratic polities perhaps during debates abound funding or other ongoing institutional issues. Indeed, the liberal Moravcsik (1997, p. 540) has posited a 'liberal-constructivism' to explain exactly this situation. In a given context, 'ideas and communication matter when they are most congruent with existing domestic values and institutions'.[10] Moravcsik's contention explains empirical results like those obtained by Lumsdaine (1993, p. 273) who found that foreign assistance programs grew out of 'ideas and values drawn from the general ethical traditions and the domestic political experience' of donor countries.

Conceivably, a liberal-constructivist synthesis untangles the puzzle of 'democratisation' in global governance, explaining the pervasiveness of demands for democratic norms despite the obvious indifference to such principles in very many states of the Global South. A synthesis also buttresses the initial argument about universal membership since the idea that affected parties should have a voice in their own governance is a widely appealing democratic ideal. In practice, liberal-constructivism means that a very large community of states and non-state actors potentially agree that international organisations should be made more transparent and broadly participatory to assert legitimate authority.

THE GLOBAL ENVIRONMENT FACILITY

In this section of the chapter, the Global Environment Facility is carefully scrutinised to understand the importance of legitimacy in an important debate about international structures. The following discussion focuses on the GEF's procedures since the two UNCED conventions specifically required institutional reforms involving universal state membership, democracy, and transparency. However, it is worth noting before proceeding that substantive legitimacy questions, which will not be addressed here, were also raised in the negotiation about GEF restructuring. Many states and transnational NGOs forcefully attacked the entity for focusing on long-range global environmental concerns like climate change and ozone depletion while explicitly ignoring the South's pressing needs for basic sanitation, clean air, and uncontaminated drinking water (Payne, 1998; GEF-NGO Network, 1998). These discussions caused the GEF to add land degradation to its list of focal areas, but substantive issues were generally not on the final agenda when states were restructuring the GEF (Silard, 1995, p. 633).

In the first section to follow, the illegitimacy of the pilot GEF is explained, with an emphasis placed on the structural and procedural criticisms levied by the Global South and various concerned transnational NGOs. Then, the second section explores the reconstructed GEF and shows how the most important structural changes explicitly sought to address the perceived legitimacy failings. In the chapter's conclusion, the ongoing and future propriety of GEF authority is briefly assessed in light of both its promise and shortcomings.

The Illegitimacy of the Pilot GEF

Developing states and a profusion of sympathetic NGOs considered the original GEF an illegitimate institution. Indeed, process issues were from the very beginning quite important to these opponents. The most significant condemnation emanated from the Global South. In the context of GEF restructuring negotiations, developing states demanded the fundamental reforms that were later echoed by the majority at UNCED: greater transparency, democracy, and universality in membership (Gupta, 1995, p. 28). Additionally, even prior to the Earth Summit, a group of 40 influential NGOs from all over the world produced a likeminded statement about GEF governance for an upcoming Participants' Meeting in Geneva. In that document (NGO Statement, 1993, p. 261), NGOs called for 'transparency, a participatory decision-making process and public accountability'.

Most interestingly, these arguments seemed to have broad appeal as they were soon repeated even by governments of important donor nations. For example, at a 1993 congressional hearing on GEF funding, a United States (US) Treasury Department official (Summers, 1993, p. 73) stated that 'a restructured GEF must be transparent and accountable, and must provide for the active participation of affected peoples and NGOs'. Similarly, British government officials reportedly shared many of these views (Tickell, 1994). The NGO Statement (1993, p. 261), however, noted that while 'there is a certain consensus on these principles' it existed only on a 'theoretical and rhetorical level'. The challenge was to institutionalise the shared understandings into GEF's practices.

To begin, it was fairly easy for critics to highlight the fact that the GEF did not encompass the universe of states. The original entity simply did not have anything like universal state membership. In fact, only 29 nations joined the pilot GEF. Worse, only a dozen developing countries became members of the experimental institution in its first three years of operation. Ordinarily, as Gupta (1995, p. 39) points out, there is great temptation for impoverished states to accept financial assistance regardless of the specific circumstances. Since the GEF was created principally to provide assistance to these states, the failure of affluent nations to convince a wide array of potential patrons to enlist highlighted basic institutional illegitimacy.

The argument about the GEF's lack of democracy was a bit more complicated, but it was certainly not obscure. Democracy, virtually all the relevant actors agreed, 'implies equity, or balance between donors' and recipients' views' (Mertens, 1994, p. 106). However, the Global South clearly perceived the GEF in much the same way as it did other Bretton Woods organisations. Just like the World Bank or the International Monetary Fund, the GEF was an inequitable instrument dominated by rich donor nations and almost exclusively reflective of their interests. In many ways then, debate about the GEF merely provided a new opportunity to rehash longstanding North-South disputes about their 'structural conflict' (Krasner, 1985).

Procedurally, affluent states favour multilateral development institutions featuring weighted voting mechanisms common to the World Bank and IMF, while states of the South seek one-state, one vote procedures used in the UN General Assembly. The South and sympathetic NGOs convincingly label the latter processes 'democratic,' but of course both North and South want to be able to control decisions regardless of the terminology. Schemes which weight votes according to financial contributions invariably provide a relatively small number of affluent states with the overwhelming balance of authority, while designs that use equal voting rights easily favour the Global South since an overwhelming majority of the world's states are poor. Because rich donor nations wrote the original GEF guidelines, the World Bank was selected to make virtually all its important decisions. As pilot-era GEF administrator Ian Johnson (quoted in Tickell and Hildyard, 1992, p. 82) proclaimed, 'In the GEF, the World Bank is judge, jury and executioner'. Daily financial management was exerted by the Bank, it was named the Administrative agency, and the GEF was physically located on Bank premises. As a result of these stipulations and others, 'the GEF structure was resoundingly rejected by the majority of developing countries' (Werksman, 1995, p. 51). Quite simply, they viewed it as an illegitimate institution.

Furthermore, NGOs argued that the GEF was undemocratic because it excluded their voices from important decisions. Since the World Commission on Environment and Development (known as the Brundtland Commission, 1987) issued its seminal report, sustainable development advocates have often framed arguments around the need for widespread participation from various local and global actors. A broad call for participation can be found, for example, in Agenda 21. Even in the specific context of the pilot-era GEF, both rich and poor states agreed that NGOs were quite valuable in providing advice about the design of projects and were perhaps crucial to implementing them successfully (Fairman, 1996, p. 61). More broadly, Raustiala (1997b) argues that states increasingly recognise the virtues of NGO participation (they provide otherwise unavailable information, for example) and are thus now commonly incorporating inclusive language in environmental accords and institutions.

In the pilot phase GEF, however, direct 'NGO participation was severely restricted' (Rowlands, 1995, p. 204). While often consulted about projects and briefed about meetings, NGOs were excluded from the Participants' Assembly and from Implementation Committee deliberations (Reed, 1993, p. 205; Jordan, 1994, p. 32). In short, they were denied a meaningful voice in decision making. Critics also charged that affected peoples in the South, often represented by local grassroots NGOs, were also frequently denied direct input into the GEF project cycle. In the words of a former NGO representative (Reed, 1993, p. 204) who had sustained interaction with the pilot stage agency, 'dissatisfaction with the GEF ... runs very deep in the NGO community both in the North and South'. For these reasons, numerous activist NGOs refused to support the pilot-stage GEF and denounced it instead as a 'green menace' (Tickell and Hildyard, 1992, p. 82; Fairman, 1996, pp. 65-6; Rich, 1994). As Raustiala

(1997) documents, numerous states, based in part on their own participatory experience with domestic environmental regulation and impact assessment, share this NGO perspective. Indeed, they increasingly recognise a 'norm of participation' in international environmental law and this norm is having a revolutionary effect on the practices of many institutions. In any event, because of its restrictive guidelines, the GEF was not viewed as legitimate by a large number of NGOs and a growing throng of sympathetic states.

Finally, both states of the Global South and transnational NGOs criticised the pilot-era GEF for its secrecy. As the World Bank often did, the GEF denied NGOs access to all but very basic documents about projects (Rich, 1994, pp. 176-8). This made it quite difficult both to monitor and evaluate projects and to provide meaningful inputs into decision making. However, despite the importance of information, it is unclear if the GEF's lack of transparency alone could have proven sufficiently compelling to undermine its institutional legitimacy. While negotiators recognised that failure to achieve universal membership and democratic decision making might have killed the GEF's future, the issue of transparency was implicitly tied to the question of NGO participation. As Florini (1998, p. 62) argues, transparency encourages 'devolution' of authority 'from government to civil society'. Still, Florini (1998, p. 50) also notes that there is a 'rapidly evolving shift of consensus among observers and actors worldwide' in favour of transparency. Critics of the GEF, like foes of the World Bank and other secretive institutions, built on this shared understanding to challenge institutional norms.

The Reconstruction of the GEF

In this section, I argue that the restructured GEF mostly addresses the concerns of the Global South and transnational NGOs regarding universal membership, democracy, and transparency. As noted above, some donor states including the U.S. seemed to embrace many of these actors' proposed reforms. The negotiations about GEF's future did not, however, simply reflect material power.

Instead, an essentially illegitimate institution was ceded proper authority only when the global community mutually agreed on dramatically transforming its procedures and practices. It was the unacceptable pilot-era entity that reflected great power interests. For the GEF to survive and succeed as a sustainable development institution, it required the support of the Global South and perhaps also the approval of transnational NGOs.

Though not quite up to UN standards, GEF membership is now nearly universal. By 1998, 164 states had joined the rebuilt institution, including over 120 states from the Global South. Thus, the GEF instrument and many of its practices are legitimated by the fact of widespread voluntary participation by sovereign equals. For example, all member states are automatically granted full voting rights in the institution's Assembly, which is supposed to meet every three years and review general policies, membership issues, and broad operational concerns. Given the numerical dominance of developing countries, it is expected (Jordan, 1994, p. 19) that the body's meetings will be democratically controlled by the Global South. The universalisation of membership, while important in legitimating institutional practices, also reflects the fact that states consider the transformed GEF a proper authority. To achieve this understanding, the institution also had to be more broadly democratised and made more transparent.

Any consideration of the democratisation of the GEF cannot focus on its Assembly. That body meets infrequently, considers only general concerns, and is not the most important within the decision-making structure. Instead, the most significant entity is the Council, which meets twice a year in Washington, D.C., controls most operational policies, and guides project implementation. GEF restructuring created a Council that features an unprecedented and arguably democratic structural arrangement. Specifically, the GEF rejected simple weighted voting mechanisms common to multilateral financial institutions and achieved a more equitable arrangement, based on a fair balance of donor and recipient authority (Harris, 1996).

World Bank Executive Directors represent states according to their financial contributions; thus, their weighted votes reflect great disparities in international wealth. A small number of affluent donors effectively control Bank decisions. By contrast, after some tough post-UNCED negotiating, GEF donor and recipient countries agreed that its governing Council would include 16 representatives from developing countries, two from economies-in-transition and 14 from developed nations.[11] In essence, it is modelled on both the UN General Assembly and the World Bank and thereby reflects key concerns of both rich and poor states. The Council has so far functioned by consensus, but technically operates under a 'double majority' voting system. All successful Council votes require both a 60 per cent majority of members and the approval of donor countries representing at least 60 per cent of contributions. In other words, the wealthiest states have agreed to share power, but they maintain an effective veto over projects and procedures they might find offensive. Indeed, the innovation might be more aptly called a 'double veto' since the South has now also been granted a mechanism for blocking projects and procedures urged on them by the North. This essentially unprecedented authority granted the South was vital in convincing them to participate in the GEF. Until this specific arrangement was agreed, it appeared as if the difficult talks on restructuring might fail altogether.

Transnational NGOs also view the GEF as legitimate now that many of their concerns about participation and transparency have been addressed. In the restructuring negotiations, NGOs were given a modicum of influence directly into the decision-making stage. Ten at one time are granted 'observer status' at GEF Council meetings, meaning in practice that five are actually in the meeting and five view proceedings from closed-circuit television. The GEF is the only multilateral financial institution that grants this status to NGOs. Moreover, since early 1995, the observing NGOs have actually been allowed to partake directly in Council meetings. The GEF grants them 'an opportunity to intervene on every agenda item' (El-Ashry, 1997).

GEF procedures stipulate that NGOs are to be consulted about many basic policy and program issues. For example, GEF implementing agencies are 'supposed to seek the advice' of NGOs 'for the identification, design, and implementation' of projects (Jordan, 1994, p. 31). Management (El-Ashry, 1997) now claims, in fact, that NGOs are already 'an important partner in the design and in the implementation of GEF projects'. In support of this assertion, the GEF-NGO Network (1998) notes that NGOs were greatly involved in the design and implementation of the Medium-Sized Grants program and have also been integrally involved in developing both operational strategies and monitoring and evaluation guidelines. NGOs are also eligible recipients of GEF grant funds. Typically, some of this cash goes to research and develop small-scale environmental projects; the nearly 1000 Small Grants Programme (SGP) projects are almost exclusively NGO-directed. However, funds for NGOs are increasingly used in the design and implementation stages of major ongoing projects. GEF Chair Mohamed T. El-Ashry (1997) reports that close to twenty percent of all funding for major projects goes to 116 different NGOs.

In regard to transparency, new procedures (Rowlands, 1995, p. 204) designate that 'all GEF-financed projects will provide for full disclosure of all non-confidential information'. Analysts argue that the mechanisms for implementing this transparency provision are genuinely 'ground-breaking' (Bowles, 1996, p. 40; Mertens, 1994, p. 106). Granting NGOs observer status at GEF Council meetings, for example, helps assure fairly open decision processes. While there are valid reasons to think that too much information might be called confidential and kept secret, the level of GEF transparency is certainly very high as compared to other multilateral financial institutions that do not have a similar presumption for information disclosure.

CONCLUSION: A LEGITIMATE FUTURE?

A great many NGOs, states, and analysts (Mertens, 1994, p. 109) consider the restructured GEF a 'vast improvement' over the pilot instrument. Nevertheless, some continue to worry about remaining 'ambiguities' in various GEF procedures. After all, the World Bank has long been criticised for its failure to live up to its most progressive guidelines. The 'Fifty years is enough' campaign was waged after many reforms had already been adopted and ignored by the Bank. That NGO effort demonstrates that the GEF's legitimacy could again be undermined if it fails to live up to its promise. Of course, some of the lingering doubts about the GEF are more important than others. For example, it is somewhat odd that the GEF-NGO Network (1998) continues to fret that the governing Council operates by consensus on 'non-issues' but uses what they pejoratively call 'modified weighted voting' on 'important substantive issues'. Ten NGO observers are allowed to watch Council proceedings and could readily challenge the GEF's claim that it has operated by consensus so far and not yet resorted to a vote (El-Ashry, 1997). Moreover, if states of the Global South found that the GEF was operating in an illegitimate manner, they could now block its operations given the 'double majority' system.

The most convincing criticisms of the restructured GEF are made by NGOs and concern their own participation and access to information. The GEF-NGO Network (1998) points out that NGO participation in formal decisions of the GEF Council has been 'very limited'. As Raustiala (1997b, p. 736) notes, the GEF includes business groups in the pool of NGOs and selects observers based on whether they have 'something specific and valuable to offer' such as information. In short, it is not at all clear that states will allow transnational NGOs to partake in decision making to the extent that the non-state actors desire. It is similarly uncertain whether the GEF will prove any more transparent than the World Bank or other secretive international institutions. As is the case with Bank loans, borrowing states have great latitude in preventing disclosure of information that they might call confidential for financial reasons (Mertens, 1994, p. 108).

Additionally, there are convincing charges by NGOs (GEF-NGO Network, 1998) that, in practice, affected peoples only participate extensively in projects during the implementation stage. Yet, various studies of grassroots involvement in development crucially recommend participation in the identification and design stages. In those phases, projects can be tailored to community needs and substantially changed, if necessary. In fact, the GEF's (1998) own study of 'Project Lessons' concluded that staff and community must interact 'especially at the outset' in order for projects to be successful. Interestingly, the GEF-NGO Network (1998) calls for a World Bank-like Inspection Panel to serve as a mechanism for resolving conflicts involving local peoples adversely affected by projects (Payne, 1996; Udall, 1998).

Despite these shortcomings, the GEF is today viewed by many (David Okali quoted in Bagla, 1998, p. 374; El-Ashry, 1998) as an institution 'open to learning from its mistakes'. While most of the ongoing debate about GEF focuses on relatively narrow practical concerns, the implications for global environment and development politics could be genuinely unprecedented. As Gupta (1995, p. 42) argues, 'restructuring of the GEF is a first step towards introducing norms and ideas that represent the views of the South ... and as the GEF becomes universal and a little bit more democratic and transparent, change becomes inevitable'. Within the GEF's domain, grassroots organisations, transnational NGOs, and impoverished states have achieved extraordinary access to an important area of global decision making and their substantive agenda clearly differs from that of the North.

Additionally, the GEF's influence could be even broader since its innovative processes and practices could be copied by other institutions. The World Bank's Assistant General Counsel (Silard, 1995, p. 653) refers to the reformed GEF as a 'state of the art' organisation that can serve as 'a model for other' institutions addressing various other global problems. Already, other development assistant entities are being recast to reflect NGO calls for participation, transparency, and accountability. This includes the US Agency for International Development (Auer, 1998), the EU (Putzel, 1998), and the World Bank (Udall, 1998). Rather than attributing change in all these institutions to the GEF process, it might be more appropriate to consider the likelihood that all these reformed institutional structures reflect newly shared democratic understandings of development.

NOTES

1. Because of these failings, and others related to its environmental record, World Bank opponents conducted a comprehensive golden anniversary campaign around the idea that 'Fifty years is enough' (Greene, 1994).

2. Anderson (1995) provides a mostly positive overview of the pilot stage GEF.

3. There is no theoretical reason to privilege states in a study of global social constructs. Even the self-proclaimed statist Wendt (1994, p. 393) explicitly discusses the need for scholars to recognise the broadening 'boundaries of political community' and to link theoretically 'the people' with transnational authorities and international institutions.

4. Agents too must be legitimate. Just as children, immigrants, or convicted criminals might be denied full participation within a polity, outlaw states or various non-state actors may also be plausibly excluded from global political communities. Moreover, states are not merely globally active agents that help construct international orders. They are also structures that must be legitimated by their constitutive domestic actors.

5. Change could also occur because of abrupt shifts in material conditions or because actors unilaterally alter practices. Even in these cases, however, some political actors most likely convinced others to behave differently.

6. Frank (1990) offers four specific standards for evaluating international legitimacy, all related more directly to law than to organisations. *Determinacy* refers to the clarity of rules as written and *adherence* means that the law has 'been made in accordance with secondary rules about rule-making' (p. 193). Additionally, *coherence* means that laws are reasonably connected to their purpose. *Symbolic validation* refers to a perceived true nexus between symbolic cues and historical, social, or political reality.

7. Held (1995) offers cosmopolitan democratic governance, grounded in transnational autonomy and obligations, as a solution to this problem.

8. Not all scholars agree about the appropriateness of these standards nor do they accept that these principles are widely embraced. Lodge (1994, p. 364) notes that international autonomy (and potentially legitimacy) can be compromised if transparency requirements decrease an institution's ability to operate effectively. Haas (1996, pp. 243-4) asserts that international-level proposals for democratic decision making, notably including transparency and participation, do not have broad global support.

9. Clark (1995, p. 508) argues that NGOs themselves claim legitimacy because of their representation of popular concerns. Raustiala (1996, p. 9; 1997, p. 573) challenges this notion since NGOs may be unaccountable and unrepresentative.

10. Though he is known for his structuralism and bracketing of domestic politics, Wendt (1992, pp. 393-4) also previously sought a bridge between constructivist and liberal theory. By contrast, Macdonald (1994) argues that extending liberal ideals to global arenas (a form of globalisation) may be undesirable as this would reproduce domestic 'conflicts and contradictions' (like class differences) as well as 'create new ones' rooted in international power disparities.

11. This was highly contentious. France, for example, reportedly threatened to slash its contributions unless parties agreed to more representatives from the Organisation for Economic Cooperation and Development (Tickell 1994).

REFERENCES

S.M. Anderson, 'Reforming International Institutions to Improve Global Environmental Relations, Agreement and Treaty Enforcement', *Hastings International and Comparative Law Review*, 18 (Summer 1995), pp. 771-821.

Anonymous, 'Designing Legitimate Institutions', *Environment* 38 (June 1996) p. 18.

M.R. Auer, 'Agency Reform as Decision Process: The Reengineering of the Agency for International Development', *Policy Sciences*, 31 (June 1998) pp. 81-105.

P. Bagla, 'Science Moves up Ladder in Push for Sustainable Growth', *Science*, 280 (17 April 1998) pp. 374-5.

M.N. Barnett, 'Bringing in the New World Order; Liberalism, Legitimacy, and the United Nations', *World Politics*, 49 (July 1997) pp. 526-51.

I.A. Bowles, 'The Global Environment Facility: New Progress on Development Bank Governance', *Environment*, 38 (April 1996) pp. 38-40.

J.T. Checkel, 'The Constructivist Turn in International Relations Theory', *World Politics*, 50 (January 1998) pp. 324-48.

A.M. Clark, 'Non-Governmental Organizations and their Influence on International Society', *Journal of International Affairs* 48 (Winter 1995) pp. 507-25.

Commission on Global Governance, *Our Global Neighborhood* (New York: Oxford University Press, 1995).

M.T. El-Ashry, 'Remarks', Speech Delivered at National Press Club, Washington, D.C., 27 May 1998 (text available at gefweb.org).

M.T. El-Ashry, 'Remarks,' Speech Delivered at NGO Consultation, Washington, D.C., 29 April 1997 (text available at gefweb.org).

Executive Summary, *Overall Performance Study of the Global Environment Facility* (1998) (available at gefweb.org).

D. Fairman, 'Commentary, The New GEF', *Environment*, vol. 36 (July/August 1994) p. 39.

D. Fairman, 'The Global Environment Facility: Haunted by the Shadow of the Future', in *Institutions for Environmental Aid*, ed. by R.O. Keohane and M.A. Levy (Cambridge: MIT Press, 1996), pp. 55-87.

M. Finnemore, *National Interests in International Society* (Ithaca: Cornell University Press, 1996).

A. Florini, 'The End of Secrecy', *Foreign Policy*, no. 111 (Summer 1998) pp. 50-63.

T.M. Franck, *The Power of Legitimacy Among Nations* (New York: Oxford University Press, 1990).

A. Gamble, 'Shaping a New World Order: Political Capacities and Policy Challenges', *Government and Opposition*, 28 (Summer 1993) pp. 325-38.

GEF, 'Summary Report of the Study of GEF Project Lessons' (January 1998) (available at gefweb.org).

GEF-NGO Network, *The GEF in the 21st Century: A Vision for Strengthening the Global Environment Facility* (26 February 1998). Available from IUCN-The World Conservation Union.

L. Gordenker and T.G. Weiss, 'NGO Participation in the International Policy Process', in Weiss and Gordenker (eds) *NGOs, The UN, and Global Governance* (Boulder: Lynne Rienner, 1996), pp. 209-21.

S.G. Greene, 'Reforming Global Institutions', *Chronicle of Philanthropy,* vol. 7 (13 December 1994) pp. 1, 7-10.

J. Gupta, 'The Global Environment Facility in its North-South Context,' *Environmental Politics*, vol. 4 (Spring 1995) pp. 19-43.

P. Haas, 'Is 'Sustainable Development' Politically Sustainable?' *Brown Journal of World Affairs*, vol. 3 (Summer/Fall 1996) pp. 239-47.

P.G. Harris, 'Considerations of Equity and International Environmental Institutions', *Environmental Politics*, vol. 5 (Summer 1996) pp. 274-301.

D. Held, *Democracy and the Global Order, From the Modern State to Cosmopolitan Governance* (Stanford: Stanford University Press, 1995).

A. Jordan, 'Paying the Incremental Costs of Global Environmental Protection: The Evolving Role of GEF', *Environment*, vol. 36 (July/August 1994) pp. 12-20, 31-6.

A. Jordan, 'Designing New International Organizations: A Note on the Structure and Operation of the Global Environment Facility', *Public Administration*, vol. 73 (Summer 1995) 303-12.

M.E. Keck and K. Sikkink, *Activists Beyond Borders, Advocacy Networks in International Politics* (Ithaca: Cornell University Press, 1998).

R. Koslowski and F.V. Kratochwil, 'Understanding change in International Politics: the Soviet Empire's Demise and the International System', *International Organization* 48 (Spring 1994) pp. 215-47.

S.D. Krasner, *Structural Conflict, The Third World Against Global Liberalism* (Berkeley, CA: University of California Press, 1985).

R.D. Lipschutz, 'Reconstructing World Politics: The Emergence of Global Civil Society', *Millennium* 21 (Winter 1992) pp. 389-420.

K.T. Litfin, *Ozone Discourses, Science and Politics in Global Environmental Cooperation* (New York: Columbia University Press, 1994).

J. Lodge, 'Transparency and Democratic Legitimacy', *Journal of Common Market Studies* 32 (September 1994) pp. 343-68.

D.H. Lumsdaine, *Moral Vision in International Politics, The Foreign Aid Regime 1949-1989* (Princeton: Princeton University Press, 1993).

L. Macdonald, 'Globalising Civil Society: Interpreting International NGOs in Central America', *Millennium* 23 (Summer 1994) pp. 267-85.

S.K. Mertens, 'Towards Accountability in the Restructured Global Environment Facility', *Review of the European Community and International Environmental Law* 3 (1994) pp. 105-10.

A. Moravcsik, 'Taking Preferences Seriously: A Liberal Theory of International Politics,' *International Organization* 51 (Autumn 1997) pp. 513-53.

NGO Statement to the Participants Meeting in Geneva, 14 February 1992, 'GEF Governance' *Authorizing Contributions to IDA, GEF, and ADF* 103rd US Congress, first session, House of Representatives, Subcommittee Hearing of the Committee on Banking, Finance and Urban Affairs (5 May 1993) pp. 257-62.

R.A. Payne, 'Deliberating Global Environmental Politics', *Journal of Peace Research* 33 (May 1996) pp. 129-36.

R.A. Payne, 'The Limits and Promise of Environmental Conflict Prevention: The Case of the GEF', *Journal of Peace Research* 35 (May 1998) pp. 363-80.

J. Putzel, 'The Business of Aid: Transparency and Accountability in European Union Development Assistance', *Journal of Development Studies* 34 (February 1998) pp. 71-96.

K. Raustiala, 'Democracy, Sovereignty, and the Slow Pace of International Negotiations,' *International Environmental Affairs* 8 (1996) pp. 3-15.

K. Raustiala, 'The "Participatory Revolution" in International Environmental Law', *Harvard Environmental Law Review* 21 (1997) pp. 537-86.

K. Raustiala, 'States, NGOs and International Environmental Institutions', *International Studies Quarterly* 47 (December 1997b) pp. 719-40.

D. Reed, 'The Global Environment Facility and Non-Governmental Organizations', *American University Journal of International Law and Policy* 9 (1993) pp. 191-213.

C. Reus-Smit, 'The Constitutional Structure of International Society and the Nature of Fundamental Institutions,' *International Organization* 51 (Autumn 1997) pp. 555-89.

B. Rich, *Mortgaging the Earth, The World Bank, Environmental Impoverishment and the Crisis of Development* (Boston, MA: Beacon, 1994).

T. Risse-Kappen, 'Ideas do not Float Freely:Transnational Coalitions, Domestic Structures, and the End of the Cold War', *International Organization* 48 (Spring 1994) pp. 185-214.

T. Risse-Kappen, 'Collective Identity in a Democratic Community: The Case of NATO', in P.J. Katzenstein (ed.) *The Culture of National Security; Norms and Identity in World Politics* (New York: Columbia University Press, 1996) pp. 357-99.

I.H. Rowlands, *The politics of global atmospheric change* (New York: Manchester University Press, 1995).

N.H. Samhat, 'International Regimes as Political Community', *Millennium* 26 (1997) pp. 349-78.

S.D. Sharma, 'Building Effective International Environmental Regimes: The Case of the Global Environment Facility', *Journal of Environment and Development* 5 (March 1996) 73-86.

K. Sikkink, 'Human rights, principled issue-networks, and sovereignty in Latin America', *International Organization* 47 (Summer 1993) pp. 411-41.

S.A. Silard, 'The Global Environment Facility: A New Development in International Law and Organization', *George Washington Journal of International Law and Economics* 28 (1995) pp. 607-54.

P.J. Spiro, 'New Global Communities: Nongovernmental Organizations in International Decision-Making Institutions', *The Washington Quarterly* 18 (Winter 1995) pp. 45-56.

L. Summers, 'Prepared Statement,' *Authorizing Contributions to IDA, GEF, and ADF* 103rd US Congress, first session, House of Representatives, Subcommittee Hearing of the Committee on Banking, Finance and Urban Affairs (5 May 1993) pp. 63-81.

L.E. Susskind, *Environmental Diplomacy, Negotiating More Effective Global Agreements* (New York: Oxford University Press, 1994).

O. Tickell and N. Hildyard, 'Green Dollars, Green Menace', *The Ecologist* 22 (May/June 1992) pp. 82-3.

O. Tickell, 'Funding Dispute Could Hold Up Biodiversity Treaty', *Nature* 367 (27 January 1994) p. 309.

L. Udall, 'The World Bank and Public Accountability: Has Anything Changed?' in J.A. Fox and L.D. Brown (eds) *The Struggle for Accountability, The World Bank, NGOs and Grassroots Movements* (Cambridge: MIT Press, 1998), pp. 391-436.

N. Van Praag, 'The Global Environment Facility, Instrument Establishing', *International Legal Materials* 33 (September 1994) pp. 273+ (document available at www.dc.enews.com/magazines/ilm/archive).

A.E. Wendt, 'The agent-structure problem in international relations theory', *International Organization* 41 (Summer 1987) pp. 335-70.

A.E. Wendt, 'Anarchy is What States Make of It: The Social Construction of Power Politics', *International Organization* 46 (1992) pp. 391-425.

A.E. Wendt, 'Collective Identity Formation and the International State', *American Political Science Review* 88 (June 1994) pp. 384-96.

J. Werksman, 'Consolidating Governance of the Global Commons: Insights from the Global Environment Facility,' *Yearbook of International Environmental Law* 6 (1995) pp. 27-63.

World Commission on Environment and Development, *Our Common Future* (New York: Oxford University Press, 1987).

Z. Young and S. Boehmer-Christiansen, 'The Global Environment Facility: An Institutional Innovation in Need of Guidance?' *Environmental Politics* 6 (Spring 1997) pp. 193-201.

7 Encouraging Participation in International Environmental Agreements

Elizabeth R. DeSombre

Most international environmental problems require widespread cooperation if they are to be addressed successfully. Emissions of substances that harm the ozone layer, create acid rain, or contribute to global climate change come from a large number of states. Even if one or several try to limit their output of these substances, the broader problem will not be solved unless most states that emit these substances do so as well. Similarly, the protection of international or transboundary resources, like fisheries or endangered species, requires that all major consumers of these resources limit their consumption; otherwise, the action by one or more who have not agreed to do so can undermine the ability to protect the resource.

The inherent benefits of international environmental agreements are sufficient to induce many states to join them; indeed it is because there are benefits to these agreements that states create them in the first place. Collaboration in scientific research produces better information. Group action to mitigate environmental destruction can accomplish more - for everyone - than individual action by some. Those who want to prevent atmospheric pollution or preserve a resource know that they accomplish that end, and thereby gain the benefits of a protected environment, more successfully by international cooperation than they could by undertaking action on their own.

Unfortunately, even states that would benefit from the protection of an environmental resource may gain some advantages from avoiding participation in international agreements designed to protect the resource in question, particularly if a sufficient number of other states do undertake such action. In a situation with prisoner's dilemma or tragedy-of-the-commons incentives, a state's second best option is to protect the resource collectively, but it would be better off if it could be a free rider on everyone else's protection efforts. If a sufficient number of states attempt to realise the gains from free riding, however, such cooperation either will fall apart or will not be achieved in the first place. Efforts to encourage participation by relevant states are therefore essential ingredients in international environmental agreements.

Aside from advocating the simple environmental benefits of international cooperation, then, states and international organisations have developed a wide variety of tools to encourage participation in international environmental agreements. Economic sanctions, undertaken either unilaterally or as part of a treaty, have been used to bring states into agreements on whaling, fisheries, protection of the ozone layer, and trade in endangered species, to name a few. Funding mechanisms, either private, such as when NGOs or states pay the membership and travel costs for other states to join the whaling agreement, or within the context of international agreements as in those to protect the ozone layer, conserve biodiversity, and prevent global climate change, have recently been used to entice states into environmental agreements. Differential obligations for those otherwise reluctant to participate are also becoming more prominent, as can be seen in agreements to protect the ozone layer and mitigate global climate change. The creation of club or private goods, such as access to information or other benefits of participation, has had an impact on the behaviour of states in preventing intentional oil pollution at sea and in bringing states into agreements relating to shipping practices.

It is important to examine the implications of these types of participation mechanisms. What do we know about the conditions under which economic sanctions, funding mechanisms, differential obligations, or the creation of private goods have worked generally to persuade states to take action they might otherwise resist? Are there any disadvantages to using these types of mechanisms, or conditions under which they are likely to be more or less successful? This chapter examines these types of mechanisms both generally and in the context of specific environmental illustrations, to determine the costs and benefits of these methods for influencing participation in international environmental agreements.

ECONOMIC SANCTIONS

Theory

The threat and imposition of economic harm has long been a foreign policy tool used to persuade states to change their behaviour. The usefulness of sanctions has been analysed within a number of different policy issues, from overthrow governments to compensation for expropriated industries or support of civil and political rights. Although persuading states to join environmental treaties is likely to meet with far less resistance than some of the above-mentioned targets of sanctioning efforts, the study of the effectiveness of sanctions in general can be instructive in looking at the implications of using this method of persuasion within the context of environmental treaties.

The first issue discussed about sanctions is whether they accomplish their stated goals; indeed, some argue that they are rarely intended to change the behaviour of foreign states but rather are meant to appease the domestic population of the sending state (Hufbauer, Schott, and Elliott, 1990, p. 11; Malloy, 1992). Influential theorists have declared that sanctions do not succeed in persuading states to change their policies (Doxey, 1987; Knorr and Trager, 1977), or can even be counterproductive (Galtung, 1967). Even Hufbauer, Schott and Elliott, who began their study with a goal of rehabilitating the idea of sanctions, ultimately conclude that 'although it is not true that sanctions 'never work,' they are of limited utility in achieving foreign policy goals that depend on compelling the target country to take actions it stoutly resists' (1990, p. 92).

To the extent that sanctions do work, what factors determine how successful they will be? The credibility of the threat is important; states are unlikely to respond to threats they do not believe will be implemented. There may be certain elements of the way a threat is constructed that will make it more or less likely to be credible. The extent to which a threat, if imposed, will hurt the sending state as well as the target state increases the difficulty in imposing it and therefore decreases its credibility (Krasner, 1977, p. 160). Characteristics of the target states may also be important – how vulnerable are they to whatever threat is being made? Vulnerability may have to do with general characteristics of the health and stability of a state (Hufbauer, Schott, and Elliott, 1990, pp. 46, 97-8), or more specifically how much the state depends on trading the specific goods the sanctions are targeted against (Doxey, 1987, pp. 110-11; DeSombre, 2000).

Another important concern that arises with use of sanctions is legitimacy. Is it acceptable to threaten or impose economic harm to convince others to take some desired action? It is argued that using methods considered coercive or illegal, particularly trade sanctions, may be counterproductive. States may resist such pressure and may also retaliate (Bhagwati, 1990, pp. 33-35). There is some concern as well that the use of sanctions, particularly outside an agreement, can undermine the agreement itself (Charnovitz, n.d., p. 4). Do concerns for the legitimacy of sanctions either lessen their frequency or their success?

Practice

Empirically, economic sanctions have been used both inside and outside treaties in support of the goals of environmental agreements. In particular, there is a long history of unilateral trade measures used to persuade states to take a desired environmental action, and some of these have been directed specifically at convincing states to join treaties. The United States has been the biggest practitioner of this method. It has threatened states with economic sanctions to persuade them to join or accept decisions of the governing organisations of the International Commission for the Northwest Atlantic Salmon Fisheries (ICNAF), the International Convention for the Regulation of Whaling, the Convention on International Trade in Endangered Species of Fauna and Flora (CITES), the Inter-American Tropical Tuna Convention (IATTC) and the International Convention for the Conservation of Atlantic Tunas (ICCAT).

A major use by the United States of threats of trade restrictions was to induce states to join or accept all the provisions of environmental treaties pertaining to the International Convention for the Regulation of Whaling. U.S. legislation allowed restriction of imports of fishery products from states that 'diminish the effectiveness' of the International Whaling Commission's policies. The U.S. officially threatened (though never sanctioned) a number of whaling states for not participating fully in the agreement. Early measures were threatened against states like Chile, South Korea, Peru, and Taiwan who had not signed the treaty but were nevertheless catching whales. All of these, except Taiwan, acceded to the treaty after trade restrictions were threatened. Taiwan stopped all foreign whaling and then agreed to stop whaling altogether, so it surpassed the obligations it would have undertaken had it joined.

Trade restrictions were also threatened against states that had signed the treaty but refused to comply with all its provisions. Iceland, Japan, South Korea (once it had joined), Norway, Spain, and the then Soviet Union were all targeted on this basis. Early in the process these threatened sanctions had some degree of success. Iceland repeatedly improved its policies relating to scientific whaling. Japan withdrew several of its objections, including the objection to the moratorium on commercial whaling. South Korea also withdrew several objections and suspended scientific whaling. The USSR and Norway removed some of their objections to policies and agreed to uphold the moratorium on commercial whaling (DeSombre, 2000, pp. 209-212). More recently, however, these threats of sanctions have not had their desired effect. Norway resumed commercial whaling (legally within the agreement, because it had never dropped its earlier objection). Iceland also withdrew from the agreement in order to preserve its right to hunt whales. Japan likewise refused to modify its scientific whaling program, despite being threatened with U.S. sanctions.

The U.S. also threatened sanctions against states that 'diminish the effectiveness' of international programs to protect endangered species (United States Public Law 95-376). Some of these threats were unilateral and outside of the auspices of the international agreements they were designed to support. For instance, the United States imposed sanctions on wildlife imports from Singapore in 1986, due to its refusal to adhere to international protection measures for wildlife. Singapore was not a CITES member and was not fulfilling CITES obligations. In 1987, shortly after the imposition of sanctions, Singapore acceded to CITES (*Federal Register*, 1986, pp. 34159, 36864, 47064). Japan, which was a CITES member, had opted out of CITES provisions that would require it to protect sea turtles. The U.S. threatened Japan with sanctions in 1991, after which Japan agreed to remove its reservation to the protection of several species of sea turtle (Lancaster, 1991, p. A3).

The U.S. also used economic sanctions in support of international agreements relating to tuna. The Tuna Conventions Act notes that 'the achievement of the conservation objectives . . . is dependent upon international cooperative efforts to implement the Commission's recommendations' (*United States Code of Federal Regulations* 50, 281.2 (d)). This includes those who are not parties to the agreements (the Inter-American Tropical Tuna Convention and the International Convention for the Conservation of Atlantic Tunas). The Act authorises sanctions against states that do not uphold the Commission's conservation recommendations. According to the House Report on the bill, the power to impose sanctions is 'intended to encourage effective cooperation with the conservation program by countries which are not parties to the conventions' (*United States Code Congressional and Administrative News*, 1971, p. 2409). The U.S. imposed sanctions under this legislation against Spain, and threatened sanctions against Panama, in 1975. The U.S. determined that these two states were fishing for tuna in the regulatory area and were not members of the IATTC. Panama made a commitment to enforce the agreement for its flagships, so the U.S. did not impose sanctions. Spain, however, refused to join the agreement or follow its regulations, so yellow fin tuna from Spain was denied entry into the US (*United States Federal Register*, 1975, p. 48160). These restrictions remained in effect until 1983 when the IATTC stopped issuing catch regulations (*United States Federal Register*, 1983, p. 32832).

Trade sanctions have also been used within agreements themselves to give states the incentive to join. One of the earliest uses of trade restrictions to gain participation was in the Convention for the Preservation and Protection of Fur Seals (1911) among Great Britain, Japan, Russia, and the United States. In addition to establishing measures for seal conservation, the treaty obligated the signatories to refuse to import seals and other regulated species from those states that caught seals in a manner prohibited by the treaty (Charnovitz, 1993, p. 40). This provision was designed to make it less attractive for states that engaged in the fur trade to remain outside the treaty.

The major treaties that require sanctions include the Montreal Protocol on Substances that Deplete the Ozone Layer and the Convention on International Trade in Endangered Species of Wild Fauna and Flora (CITES). The Montreal Protocol (1987) has perhaps one of the best known multilateral policies of trade restrictions for the purpose of encouraging participation in the agreement. Article 4 of the Protocol prohibits trade in controlled substances (as well as products containing controlled substances) between parties and non-parties. The one exception is that states that are not members of the treaty but have nevertheless fulfilled all treaty obligations are allowed to engage in trade with member states (Brack, 1996, p. 50).

This policy was likely to be highly effective once the major producers of ozone-depleting substances became members. Other users would have to join as well (or perform the same obligations required of members) to gain access to the regulated products. This would be important since CFCs were initially not to be phased out completely, and they continue to be important industrial substances in developing countries. A number of non-member states, such as Malta, Jordan, Poland, and Turkey applied to be allowed to trade, while meeting the obligations, during the period that they had not ratified the agreement. Some states have cited the trade restrictions as a major reason for joining the agreement. In particular, South Korea did not initially join the Protocol because it did not believe that it would be adequately compensated for the costs of its phase-out activities. But the threat of trade restrictions, particularly against products (like automobile air conditioners) containing CFCs, ultimately persuaded South Korea to join. Israel was also influenced by trade restrictions in its decision to join (Brack, 1997, p. 4).

CITES prohibits trade in species with non-parties unless accompanied by 'comparable documentation' that 'substantially conforms with the requirements of the present Convention' (CITES, 1973, article 10). In other words, those states that do not wish to join the Convention must still uphold its main provisions in order to trade with member states. It is unclear the extent to which CITES trade restrictions against non-parties have been directly responsible for bringing states into the agreement. Nevertheless, with the treaty-authorised sanctions described below, a number of states have been encouraged to join the agreement through fears they would be shut out of the lucrative wildlife trade.

A number of treaties either implicitly or explicitly authorise their members to impose trade sanctions to deter violations, whether those be by members or non-members. In some, the ability to take 'appropriate, legal, administrative and other measures to . . . prevent and punish conduct in contravention of the Convention (Basel Convention, 1989, Article 4)' has been written into the agreement. This authority is often believed to include the right to impose sanctions against those who contravene the convention, both members and non-members. The Convention for the Prohibition of Fishing with Long Driftnets in the South Pacific, for instance, states that parties may 'prohibit the importation of any fish or fish product, whether processed or not, which was caught using a driftnet (1989, Article 3(2)(c))'. More important are those treaty organisations that explicitly authorise their members to prohibit trade with non-member states that are hindering the effectiveness of the treaty by their non-participation. Those that authorise them include CITES and the ICCAT. In both of these cases, the primary state to threaten or impose sanctions has been the United States, though Japan also was involved in the ICCAT sanctions.

In addition to the CITES restrictions on trade with non-member states, its governing organisation has the authority to call for sanctions against states that are not upholding its regulations, including non-member states. For instance, in 1987 the Conference of the Parties called on members to 'use all possible means (including economic, diplomatic and political) to exert pressure on countries continuing to allow illegal trade in ivory, in particular Burundi and the United Arab Emirates' (CITES, 1987). Burundi joined the agreement in 1988 and the UAE, which was previously a member but withdrew from the treaty in 1987, rejoined in 1990. In 1991, the CITES Standing Committee recommended that members ban the trade of restricted species with Thailand because it was not preventing trade in protected species (Charnovitz, 1993, p. 27). Although Thailand signed the Convention in 1983, it had not passed legislation to implement it (Abramson, 1991, p. A9). A number of CITES member states refused to accept imports of Thai wildlife, most notably, orchids. Shortly after sanctions were imposed, Thailand passed implementing legislation (French, 1995, p. 23; Stier, 1991, p. 6A); sanctions were lifted a year later. The CITES standing committee warned in 1993 that South Korea, China, Taiwan, and Yemen, the latter two of which are not CITES members, would be subject to sanctions if they did not implement restrictions on trade in endangered species. Officials from Yemen indicated that country's intention to ban trade in rhino horn, and use substitutes for the making of ceremonial dagger handles (Press, 1993, p. 12). Since Yemen gave indications that it would fulfil the requirements of membership (despite not signing the treaty), sanctions were not imposed. South Korea also did not receive sanctions, probably for political reasons. CITES recommended sanctions against China and Taiwan (the latter of which is not allowed to join CITES because it is not a member of the United Nations), for general inaction in upholding CITES regulations, particularly those relating to trade in rhinoceros and tiger parts. The United States threatened trade restrictions on both states but imposed them only on Taiwan, on 2 August 1995. Both China and Taiwan agreed to take steps to improve their enforcement of CITES regulations, and the U.S. lifted the sanctions against Taiwan on 30 April 1997 (UPI Wire Report, 1994; *United States Federal Register*, 1997, p. 23497).

The International Convention for the Conservation of Atlantic Tunas has also experienced problems with participation, both in terms of persuading member states to uphold the recommendations of its commission and in bringing non-member states, fishing in the regulated area, into the agreement. In 1994 the organisation, at the prompting of Japan and the United States, adopted the 'Action Plan to Ensure the Effectiveness of the Conservation Program for the Atlantic Blue fin Tuna'. Under this plan, the organisation first identifies those that have been fishing in the regulated area in a way that 'diminishes the effectiveness' of ICCAT regulations, then contacts the non-member states to request that they adhere to ICCAT's regulations. If, within a year of being contacted by ICCAT, the non-party has not taken steps to adhere to the organisation's policies, the Commission is to recommend to member states that they prohibit imports of blue fin tuna by the identified state. This plan was first implemented beginning in 1995, when Belize, Honduras, and Panama were identified as non-member states that were fishing in the Mediterranean during a closed season. The Commission requested that these states change their behaviour. When, after a year, these states had not joined the treaty or brought their fishing practices into line with it, the ICCAT Commission requested that members prohibit imports of blue fin tuna from these countries (Carr, 1997). The United States prohibited blue fin tuna imports from Belize and Honduras as of 20 August 1997 and from Panama as of 1 January 1998 (*United States Federal Register*, 1997, pp. 44422-3). Japan, which imports at least 60% of the world's blue fin tuna catch, also imposed sanctions, lessening its total imports of the fish by 10% (AP Worldstream, 1996). Although it is too soon to know the final results of these sanctions, there are preliminary signs that they are having an effect - an indication of the efficacy of such threats. Panama, which initially denied violating international regulations (Reuters World Service, 1996), indicated that it would take greater steps to prevent its ships from catching tuna in violation of ICCAT policies (Grant, 1996). Belize also said it would attempt to curb acts by its fishers that violate ICCAT rules (Tighe, 1996).

Lessons

The use or threat of sanctions to encourage states to join environmental treaties works more frequently than those who study sanctions in general suggest. In most of the cases discussed here, states either joined the treaty when threatened with sanctions, or aligned their behaviour with treaty requirements without joining. What is it about these types of sanctions that makes them more effective than the general experience with sanctions indicates? In the first place, the demands are often modest. Although joining an environmental treaty may require costly action, it is rarely as difficult for a state to limit its fishing activities or cease trade in endangered species than to change its form of government or provide political rights it has previously denied. Frequently, becoming a treaty member is associated with some benefits as well. Some of the more direct side payments may benefit states that join, but the environmental advantages of protecting a valuable resource, even if insufficient to induce the state to join initially, may also make giving in to threats less problematic than in other issue areas.

Second, the threats of sanctions in environment cases are particularly credible. Most of the time the sending states do not lose much and may, in fact, gain from the process of imposing sanctions. They are therefore quite willing to follow through on their threat, unlike in situations where the U.S. will lose an important market for its exports if it carries out its threat. In the unilateral cases examined here, the sanctions are almost always import restrictions, and the imposition of sanctions is supported by domestic industries that would prefer that competitor goods from elsewhere (particularly goods produced more cheaply because they do not have to meet environmental standards) be kept out. In some cases, the sanctions threatened are designed to protect those involved in the treaty from suffering the disadvantages of free ridership as much as actually bringing others into the agreement (DeSombre, 1995). This incentive makes it even more likely that threatened sanctions would be imposed.

The same is true to some extent with multilateral sanctions or treaty-based sanctions. Most prevent the non-member state from exporting to others that have joined the treaty. Moreover, industry actors in the member states may benefit from the imposition of these sanctions. For that reason, a threat that sanctions will be imposed can be quite credible, and is likely to be taken more seriously than threats of other types of sanctions. The treaty-based sanctions that are not import restrictions, such as the blanket ban on trade in controlled substances under the Montreal Protocol or trade restrictions in wildlife under CITES, may also have a high level of credibility. These types of restrictions, although they target both imports and exports, rely on the policy of an international organisation, rather than the decisions of individual states. Sending states are much more likely, therefore, to follow through with threatened sanctions in this issue area than in others.

The importance of credibility can be seen in recent efforts to threaten states to change their whaling behaviour. Although initial threats were sufficient to bring about great changes, later threats have failed to accomplish their goals. It is possible that the fact that threatened sanctions have never been imposed within this issue area decreases their credibility. For threats to be effective, target states must believe that the threat will be carried out if they do not take the action requested. If, when a threat is made, the target state does not respond, and the threat is not carried out, that state learns that it may not suffer for refusing to change its policy. Furthermore, others that see the threat not implemented may question the credibility of later threats made towards them.

Credibility is not the only factor, however. In some cases, states refuse to join an agreement even when sanctions are imposed. Spain's refusal to join the IATTC is one example. It is likely that the cost of the sanctions, both in the short term and in the long-run, then becomes important. For Spain, fishing is a crucial industry; the cost of losing the U.S. as an export market is probably less important than restricting its fishing practices would be. This is supported by the multiple instances in which Spain has avoided undertaking fishing conservation measures. Likewise, states like Norway that do not rely heavily on fish exports to the U.S. are not likely to be persuaded, by even the most credible sanctions, to comply with the International Whaling Commission.

The whaling illustration also reveals the potential for threats of sanctions to undermine an agreement. In a normal sanctions process, once states have come into an agreement, they frequently discover that their obligations are not as onerous as they expected, or they change their behaviour such that it becomes more costly for them to stop their participation than continue it. That process has not happened with the sanctions used to entice states into the whaling agreements. There is currently a crisis within the agreement, with some states withdrawing, others engaging in commercial whaling despite the organisation's continuing ban, and others refusing to come into line with the IWC's approach to scientific whaling. In none of these recent cases has the threat of sanctions accomplished its stated goals. The constant use of threats, over the history of the whaling agreement, may have undermined the ability of its governing organisation to reach a consensus policy for protecting whales.

Another important but difficult question concerns the relative advantages of unilateral versus multilateral sanctions. Multilateral sanctions implemented by international organisations might, in theory, be expected to be more credible and more powerful, since they have broad support. They are also more likely to be acceptable under international trade law, and therefore might have a higher degree of legitimacy. In practice, however, unilateral sanctions (particularly those implemented by the United States) have had the most dramatic impacts in changing target state behaviour with respect to environmental issues. The examples are not easy to analyse however, since it is not possible to compare parallel cases whose differences are in the number of states imposing sanctions. While multilateral sanctions might have a greater effect in such cases, there may be other factors that decrease the likelihood that multilateral sanctions would be used at all.

ENVIRONMENTAL AID

Theory

Aid has also been a general foreign policy tool used to convince (and enable) the recipients to take actions the donors desire. The lessons learned from aid in a broader foreign policy context are likely to prove useful in examining the implications of aid as a tool for encouraging states to take on environmental treaty obligations. Aid can be seen as essentially a side-payment linked to other issues. A state's behaviour in one area (a treaty) that it may not be particularly concerned about is linked to something else (a type of aid) that it is. Using side-payments is a well-known strategy for encouraging international cooperation.

Aid can also be seen as an effort to change a state's preferences – changing the payoffs, in a game-theoretic structure. If a state's preference ordering is changed from preferring to stay out of a cooperative arrangement to preferring to join, due to aid, cooperation becomes more likely. As that example indicates, economic aid is the same type of mechanism as sanctions; they both change a state's calculation of its interests or of the costs and benefits of taking a certain action. But in terms of public perception, they are worlds apart. A target state would almost always prefer to have its cost/benefit equation changed in favour of cooperation by aid than through the removal of benefits it is already receiving. The process of providing funding to encourage states to take certain actions therefore has some advantages. It is often seen as a morally superior way to induce action. Recalcitrant states may be better able to justify their participation to their domestic populations when they receive compensation rather than threats.

The above formulation also points out potential problems with aid as a side-payment for state action. There is the possibility of moral hazard. If states only take a desired action when they receive aid, will they only continue to take that action as long as aid is provided? Will it be possible to convince them of the benefit of undertaking the action in question, or will they simply do it for the payoff? Moreover, paying states to take a beneficial action may have longer-run effects if states learn that they do not have to take action to protect the environment without receiving aid. It is conceivable that states, that might have acted on their own, will discover that they can get side-payments if they wait. This raises the cost of negotiating agreements and the likelihood that they will not be completed.

The idea of conditionality – 'the exchange of financial assistance for policy reforms' (Fairman and Ross, 1996, p. 31) – is behind much aid that is given, either explicitly or implicitly. Those who study this phenomenon offer caution, suggesting that conditionality does not frequently succeed in achieving the policy goals that accompany aid. Problems arise because once money has been disbursed there may be nothing to ensure that states live up to the promises they made when they received the aid (Fairman and Ross, 1996, p. 31). Most strikingly, scholars conclude that 'environmental aid packages will fail to persuade reluctant governments to implement policies they oppose,' especially those that are difficult (Fairman and Ross, 1996, p. 36). Aid is therefore unlikely to force a state to join an agreement. If it does, that state is unlikely to take the actions required of it within the agreement.

Practice

Within environmental issues, aid has had a wide variety of uses, with some related specifically to treaty processes. It can be directly related to the goals of a treaty, or to something else the recipient state wants, inside or outside the treaty. It can be provided by members through the treaty process itself, by states outside the agreement, or by non-state actors. It can be given as funding or as technology transfer. Although aid may be used for a wide variety of goals, even when provided through a treaty process, it is aid that is intended to bring states into agreements that is examined here. A number of international environmental agreements contain provisions for aid to those countries that agree to undertake treaty obligations. Many agreements indicate an intention to provide funding; fewer actually set up a process by which a significant amount of money is given to states.

Technology transfer is also often frequently referred to within treaties as a method to help members, particular developing countries, meet their obligations under the treaties. Explicit technology transfer as an incentive for participation in an agreement was first used in the Nuclear Non-Proliferation Treaty (Sand, 1990, p. 7; Schiff, 1984). Access to technology or technological information is promised in the Montreal Protocol, the Convention on Biological Diversity, the Framework Convention on Climate Change, and the Convention to Combat Desertification, to name only a few. Although it is included as a goal or requirement in a wide variety of environmental treaties, meaningful transfer rarely happens. The competitive position of the suppliers of technology is the sticking point. They do not want to forego profits by having the technology transferred at concessional prices. They also do not want to transfer to others the ability to make the technology itself, and they fear competition from those who can leap ahead technologically through this type of transfer (Elliott, 1998, p. 195). Technology is therefore more often transferred to developing country parties by providing financial resources that can be used to purchase technology.

Some treaties have fairly modest funding processes. The Basel Convention on the Control of Transboundary Movements of Hazardous Wastes and their Disposal, for instance, indicates that technology transfer should take place and that the Parties should consider 'establishment of appropriate funding mechanisms of a voluntary nature' (1989, Article 14). The World Heritage Convention establishes a 'World Heritage Fund' composed of compulsory and voluntary contributions from members and others, though parties may opt out of the compulsory contributions. Through the Fund, states may request financial assistance in protecting sites of important natural or cultural heritage (1989, Articles 15-26). Although the funds provided ($2.2 million annually) are not enormous, neither are the obligations, and funding may make the difference for a country between being able to protect a heritage site or having no hope of doing so. Some suggest that this fund has played an important role in bringing states into the agreement (Sand, 1990, p. 7).

The Montreal Protocol on Substances that Deplete the Ozone Layer (in the London Amendments of 1990) was the first major environmental agreement to contain provisions for aid to help parties meet the obligations they undertook when signing the treaty. Some would have joined anyway; Mexico was an early signatory to the Montreal Protocol, before the funding provisions were added. But a number of states explicitly tied their participation in the treaty process to the creation of a funding mechanism. India and China, whose participation in the agreement was essential due to their large and growing populations, rapid industrialisation, and ability to produce ozone depleting substances, refused to join the agreement unless a satisfactory aid package was designed (Benedick, 1991). The resulting funding mechanism is specified in Article 10 of the protocol (as amended), and involves funding from developed countries based on the U.S. scale of assessments, placed into a multilateral fund overseen by a committee composed both of donors and recipients (Montreal Protocol, as amended 1990, Article 10; DeSombre and Kaufmann, pp. 89-126). Once the funding mechanism was put into place, all the crucial developing countries joined the agreement.

Funding is also included in the Framework Convention on Climate Change and the Convention on Biological Diversity. Funding for both these treaties is distributed through the Global Environmental Facility, which was created at the same time as the Montreal Protocol Multilateral Fund, although it was only later that it was named the funding mechanisms for these treaties. Both treaties were negotiated as part of the United Nations Conference on Environment and Development, shortly after the London Amendments to the Montreal Protocol, and developing countries were insistent that aid be included in the agreement as a quid pro quo for signing (Dasgupta, 1994, pp. 138-9; Grubb et. al., 1993).

The Framework Convention on Climate Change requires that developed country parties provide 'new and additional financial resources to meet the full agreed costs incurred by developing country Parties' in meeting their obligations under the convention (1992, Article 4(3)). At this point, these obligations pertain to gathering and communicating information about the sources and sinks of greenhouse gases, but it is likely that any new obligations will be undertaken only with the financial assistance to meet them. The funding obligation outlined in the FCCC goes further in some ways than that under the Montreal Protocol; the convention specifies that 'the extent to which developing country Parties will effectively implement their commitments ... will depend on the effective implementation by developed country Parties of their commitments ... related to financial resources' (Article 4(7)). The Kyoto Protocol indicates further commitments for transfer of financial resources and technology (1997, Article 11).

The Convention on Biological Diversity sets forth that the Parties 'shall cooperate in providing financial and other support for ex-situ' conservation. It also addresses the transfer of technology and requires that developed country Parties provide resources to developing country Parties (1992, Articles 9, 16, and 20). As there are as yet no abatement measures required from countries that could be recipients of funding, it is currently used to create national biodiversity conservation strategies and to do inventories of biodiversity. Moreover, the CBD suggests that developing country parties will not be held to whatever obligations they have taken on if developed country parties have failed to implement their obligations to provide funding (1992, Article 20(4)).

It is also possible for aid to be given, external to a treaty, to persuade states to join. This type of aid is more difficult to examine because it may be done privately. In some cases, the aid may be minor – no more than paying the dues and travel expenses of member states. In other instances, a greater investment is made to bring states into agreements. Nevertheless, there are some examples that have come to light. We see this type of aid in the International Convention for the Regulation of Whaling. Those on different sides of the controversy over commercial whaling have persuaded additional states to join through offering some type of aid. Those who oppose commercial whaling have arguably brought such states as Antigua and Barbuda, Oman, Egypt, and Kenya into the agreement (DeSombre, forthcoming). A consultant tells of a Greenpeace plan that brought in six new anti-whaling members in the late 1970s and early 1980s by paying their dues, naming commissioners to represent them, and drafting their membership documents. For this, Greenpeace paid more than $150,000 annually (Spences, 1991, p. 174). Japanese foreign aid also arguably brought states into the agreement to vote in favour of commercial whaling. Grenada rejoined the IWC around the time that it received funding from Japan for new fishing fleets (Brown, 1993, p. 26); Dominica rejoined the IWC the year it began receiving more than $16 million in Japanese aid (Fineman, 1997, p. A1).

Lessons

Aid has thus become an important part of environmental treaties, and has undoubtedly brought states into agreements they would not have otherwise signed. When an environmental problem is low on the agenda of some states and the abatement process would be costly, aid can make the difference that persuades them to join. The Montreal Protocol would certainly not have gained the near-universal participation it has without aid, and the same is likely true of the FCCC and Convention on Biological Diversity.

Several of the caveats brought up by those who study aid more broadly are worth revisiting, however. The drive for aid has become universal enough that it is nearly impossible these days to negotiate an environmental treaty without a significant aid component. Compensation to those states that are not responsible for the bulk of the environmental harm but are being asked to undertake potentially costly steps to prevent further environmental destruction can certainly be seen as fair. But it may make the negotiation of environmental agreements more costly and therefore more difficult.

The relationship of aid to capacity brings up a related concern. While aid is certainly a noble approach for helping states join treaties that would not have the capacity to uphold them absent aid, the issue of capacity is a complicated one. Do states like India and China, prime recipients of capacity-building funding under a number of environmental treaties, really not have the *capacity* to phase out ozone depleting substances? Is it realistic to think that a state capable of building a nuclear weapon cannot find or use substitute chemicals for cleaning electronic equipment? The rush to provide aid to bring states into agreements by building their capacity might not adequately diagnose the problem.

One concern, that recipients might not live up to their commitments once they receive aid, has been avoided fairly successfully with aid provided within the context of treaties. The institutional setting makes it more likely that both sides can be held to their end of the bargain, and it generally provides a process for dealing with the situation in case that does not happen. More importantly, once a state is in an agreement, the issue of compliance and implementation can be addressed within the treaty process. The incentive not to fulfil obligations may diminish as well. In agreements dealing with issues like ozone depletion, there is reason to believe that once states begin a process of using non-ozone depleting technology their incentive to revert back to the old way of production decreases.

This concern is not completely groundless, however, from either side. In the Montreal Protocol donor states have at times been slow to contribute to the Multilateral Fund, which has slowed down the speed of implementation on the part of developing countries. The clause in the FCCC that ties performance of developing countries to donor countries living up to their financial obligations is almost certainly a response to this problem. At the same time, the Montreal Protocol provides words of caution about recipient state action. Although the aid brought states into the agreement, some of the largest developing states have used the funding while continuing to increase their use of ozone depleting substances. This is technically allowed by the agreement, but receiving funding to close down one halon plant while building another, as China did, certainly violates the spirit of the agreement.

More importantly, these concerns call into question the extent to which states that are brought into an agreement because of aid change their behaviour fundamentally, or whether they have simply been (temporarily) bought off. The fact that some of the states brought into the whaling agreement by funding from anti-whaling organisations left the treaty and later appear on lists of those brought into the agreement by pro-whaling states indicates the potential for funding to distort preferences of states. This makes the governing process within the treaty potentially more difficult. The behaviour of developing states in the Montreal Protocol, as obligations to phase out ozone depleting substances come due, will be illuminating.

DIFFERENTIAL OBLIGATIONS

Theory

Legally, all states, as sovereign, are considered to be equal in facing the international system. Many international agreements start from the assumption that all states will bear the same costs and garner the same benefits from participation. For environmental treaties in particular, the starting point is generally a set of regulations in which everyone takes the same action. But states are not equal, in terms of capabilities, interests, and willingness to undertake international regulation. When some are reluctant to take on obligations for protecting the environment, creating different types of regulations for differently situated parties can be one approach to bring states that would otherwise not join into a treaty.

In the international relations literature, selective incentives, as Peter Sand calls them (1990, p. 7), are well established as an aid to collective action. Sand points out that these types of differential obligations can paradoxically increase the level of agreement that can be reached, avoiding the least common denominator approach that often prevails when states are reluctant to agree to tough regulations.

There are a number of advantages in using differential obligations within treaties. In the first place, doing so may be considered fair. States are not realistically equal, and the environment does not respond equally to all abatement measures taken. It can be argued that the standard non-differentiated method of abatement, in which all states cut back on their emissions of controlled substances by an equal percentage, is unfair. For example, if all states have to cut emissions of a regulated substance by the same percentage, those that have a barely noticeable impact on the environment are required to take action, even though the responsibility for causing environmental damage may fall on the biggest emitters. The process of differentiating, however, may be difficult, as straying from a simple formula in which everyone takes the same action or pays the same cost may open up difficulties with negotiation.

Practice

The idea of 'common but differentiated responsibilities' has a long history within environmental treaties, though it has become more common in the last decade. Differentiation can be done on the basis of the environmental problem itself, on the capabilities of the member states, on political manoeuvring, or on some combination of all of these elements. Although there are justifications for differentiation other than enticing states into an agreement, this consideration is often primary in the discussion of which states will undertake what action. It should also be noted that there are small degrees of differentiation (the fact that under the Long-Range Transboundary Air Pollution agreement (LRTAP) what is counted as emissions for the former Soviet Union was only 'transboundary fluxes' rather than total emissions; the European Community's ability to aggregate its consumption of ozone depleting substances under the Montreal Protocol) in most treaties; what is discussed here is differentiation more broadly in the types of obligations states take on or the time at which they agree to undertake them.

Differentiation on the basis of environmental characteristics may be politically difficult, but it makes environmental sense. It may give an incentive to join to those that either receive lower abatement obligations or greater environmental benefit under such a system. LRTAP has begun to include such a calculation in its recent protocols under the idea of 'critical loads'. This plan, included in both the NO_x and second sulphur protocols, aims to reduce the amount of pollution to a level below that at which environmental damage is felt. To do so means taking into consideration characteristics of the air patterns that deliver pollutants and the susceptibility of particular areas to acidification. The two LRTAP protocols do not fully regulate based on critical loads but are attempting to move in that direction. It is not clear the extent to which this particular differentiation was done for the purpose of encouraging membership, but it will certainly beneficially influence the willingness of some member states to accept new protocols.

The Convention to Combat Desertification distinguishes member states based on environmental characteristics. The states are divided into affected parties and (implicitly) those that are not (often simply denoted as 'developed country parties'), with developed country parties charged to give support to policies undertaken by affected parties. Likewise, the International Tropical Timber Agreement differentiates states that produce and consume tropical timber in determining obligations (1983, Article 4). Although the Framework Convention on Climate Change has not formulated specific policies that involve action on the basis of environmental factors, it refers to action required to meet the needs of states likely to be harmed by climate change, such as small island states and those with low-lying coastal areas (Article 4(8)).

Differentiation on the basis of capabilities of states has also become relatively common within environmental treaties. One example is the different obligations for European Union members under the Directive on the Limitation of Emissions of Certain Pollutants into the Air from Large Combustion Plants, in which member states have different obligations depending on their economic and technical capabilities (Sand, 1990, p. 8).

Contribution of funding to support environmental treaties is almost always done with some degree of differentiation among parties based on capabilities. Generally some formula that is based on the wealth of the states in question is used. The funding for the Global Environmental Facility, the Montreal Protocol Multilateral Fund, and the Long-Range Transboundary Air Pollution agreement, to name a few, are based on this type of calculation. This determination not only parallels, but also often uses, the formula created to determine the contribution of states to the United Nations based on their income level (and capped so that no one state provides more than 25% of the funding).

Other distinctions drawn about the capabilities of states are often done at a fairly high level of generality, with states divided on the basis of income into developed and developing countries. Under the Montreal Protocol, developing country parties were initially given a 10-year grace period during which they could not only avoid meeting the phase out obligations of developed country parties, but could increase their use of ozone depleting substances, as long as they did not pass .3 kg per capita (Article 5(1)). This mechanism was explicitly intended to bring developing countries into the agreement by getting them to agree at a point when they did not yet have to meet abatement obligations (Benedick, 1991, p. 148). Stricter measures, written into the agreement, or negotiated in amendments, also contain a time delay before developing countries are responsible for implementing them. Although a few developing countries joined the protocol when initially negotiated, this provision alone was not sufficient to persuade most to join; that took the creation of the Multilateral Fund, discussed earlier.

Both the Convention on Biological Diversity and the Framework Convention on Climate Change continue the tradition of requiring a lower level of activity from developing country parties. In the CBD, developed country parties agree to contribute funding that will go to developing country parties, as described above. The FCCC indicates specific obligations for developed country parties, relating to the adoption of national policies and information dissemination, that are not required for developing countries; the Kyoto protocol requires that specific abatement measures be taken by these countries but are not required from developing countries. In addition, both the FCCC and the Kyoto Protocol add a new basis for differentiation, giving special consideration to 'countries that are undergoing the process of transition to a market economy' (FCCC, 1992, Annex I; Kyoto Protocol, 1997, Article 3(5) and Annex I). Developing (and former communist) countries refused to participate in these treaties, particularly the FCCC, if they were not given flexibility in meeting requirements. No obligations were considered for developing countries under the Kyoto Protocol, despite widespread political pressure from outside the negotiating process, including a resolution from the U.S. Senate indicating that it would refuse to ratify an agreement that did not contain obligations for developing countries.

Political manoeuvring, without a clear justification from environmental characteristics or the capabilities of member states, is also becoming more frequent as a basis for differentiation within environmental treaties. The idea of 'common but differentiated responsibilities' was taken to a new level with the Kyoto Protocol to the FCCC. In addition to providing no abatement requirements for developing countries, the 'Qualified Emission Limitation or Reduction Commitments' for developed country parties varies from 92 percent of 1990 emissions to 110 percent (Article 3(1) and Annex B). Moreover, although the Convention indicates the need to take into account 'specific national and regional development priorities, objectives, and circumstances (Article 4(1))', there is no specific formula for arriving at the differentiated commitments of the parties. Instead, the commitments were those that were politically feasible – those to which parties were willing to commit.

An earlier example of this type of political manoeuvring concerns the financial arrangements in the Convention for the Protection of the Rhine River Against Pollution by Chlorides; rather than basing the assessed duty for abatement of chloride pollution on either a state's income level or contribution to the environmental problem, the amount was based on how much the countries were willing to pay (Bernauer, 1996).

Lessons

Certain types of differentiation, particularly of obligations for developed versus developing countries, or different levels of financial obligations undertaken based on wealth, are now expected within environmental treaties, and they have played an important role in bringing states into agreements. Creating treaties that have different obligations for different sets of countries may be more complex than having uniform requirements, but under some conditions the increased level of participation may be worth the added complexity.

As differentiation of obligations within treaties becomes more common, difficulties become clear. Although basing obligations on environmental effects is the most environmentally reasonable approach to creating obligations, it has not yet succeeded politically. The most ambitious attempt to do so, using the idea of critical loads in the LRTAP second sulphur protocol, not only was profoundly watered down in the agreement but the agreement itself has not yet entered into force.

Basing differentiation on capabilities of states has gained greater acceptance and has fallen into a pattern in which developing countries as a group are treated differently from developed countries (but similarly within the group), and developed countries have financial obligations that depend on a modified form of the United Nations funding process. These arrangements are generally considered to be fair, and are reasonably stable.

What may be more problematic, however, is differentiation that is based not on the environment or capabilities, but on the political will of states. On the one hand, this type of differentiation allows for the participation of states that simply would remain outside a treaty at a higher level of obligation. To the extent that bringing states into a treaty process, even at a lower level of participation, is desirable, a case can be made for differentiating based on this principle. The difficulty is that there is no objective way to quantify political will. Additionally, differentiation based on the recalcitrance of the negotiating states, with the more stubborn ones rewarded for their efforts, opens up the possibility that states will demand individual treatment more frequently. The precedent set by the Kyoto agreement, in which states are given different abatement obligations without any effort to justify it technically (based either on the environment or on specific details of the capabilities of states) may bode ill for future negotiations.

CREATION OF CLUB GOODS

Theory

Part of the difficulty of inducing states to join international agreements comes in cases where states can receive much of the benefit from an agreement without having to join. This phenomenon happens primarily when the benefit created by cooperation is a public good or a commons resource. In those instances, the benefit is created by one or more actors, but the advantages, such as a stable global financial system or an intact ozone layer, cannot be denied to those who do not participate in providing them.

In some instances of international cooperation, however, the benefits created by an international agreement are excludable, and can be kept from those who do not participate. Not surprisingly, those types of issues create added incentives for states to join agreements. This type of issue is sometimes known as a club good, because the benefits of membership belong to those in the 'club' and not to others (Buchanan, 1965). Free trade agreements are the quintessential example of agreements that accomplish their goals in part through the creation of a set of advantages that accrue to members and are not available to those who do not join the agreement.

This phenomenon of creating a club is similar to the side payments that are the basis of environmental aid, but it has some important differences. Side payments do not require an institutionalised solution. They also are more likely to be rival than are club goods; in other words, in their purest form club goods are not diminished by the number of people that partake of the club, whereas the more states you have to pay off to join an agreement, the fewer resources there are to go around. In addition, in a situation of club goods, the club itself may be what creates the goods. Free trade, for example, does not exist without states removing trade barriers. The good provided, then, is an essential element of the cooperation, which is not the case with side payments. There are some issues that have characteristics of club goods on their own. In other cases, however, it may be possible to create elements of a club good within an international agreement.

There are important advantages to the creation of a club to provide incentives for participation in international agreements. If states can only receive certain benefits from joining, they are more likely to do so than if they will gain advantages without membership. If actions by the club members increase the benefit they receive even within the club, that benefit may increase their willingness to undertake collective action.

There are a few downsides to the idea of a club good. The most obvious one is that this type of solution will not work for many international problems. A large number of international issues are not excludable; non-members cannot be kept from the benefits of membership. The creation of a club for these issues simply is not a feasible alternative.

A second potential concern comes from the political power that can derive from exclusion. If there are some outside a given club, there is likely to be a mechanism to keep them out that may be co-opted for reasons other than increasing club membership. The U.S. insistence that China be kept out of the World Trade Organization because of its record on human rights is a perversion of what the club was intended to accomplish.

When some of the assumptions of what constitutes a club good are relaxed, there may be other difficulties with using it as a solution to addressing problems of participation. In the ideal form of a club good, as represented in the idea of free trade, the size of the club does not matter, and benefits accrue equally to all members no matter how many participate. But there may be some agreements that function like clubs in which members feel 'crowding' or 'congestion' with the addition of other participants (Cornes and Sandler, 1996, p. 348).

Practice

The creation of specific advantages that accrue only to those who are part of an agreement is not a widespread phenomenon within international environmental issues, but it can be seen within several agreements. The most interesting one is the Memorandum of Understanding on Port State Control, among a set of states attempting to uphold equipment standards from the International Convention for the Prevention of Pollution from Ships. These states created a labour-intensive system whereby they inspect ships and enter reports into a database made available to others who are part of the system. Those who use the database receive information on which ships have been inspected when, and what the results were. This increases the ability of the inspectors to stop ships that most need inspection and prevent likely pollution within their waters. The system, in effect, creates a club that gives a benefit to those who agree to undertake inspections in this manner (Mitchell, 1994).

Other issues of oil pollution, both intentional and accidental, have been addressed in ways that contain elements of the creation of club goods. When states require that tankers use particular types of equipment (double hulls to make spills less likely, or segregated ballast tanks to make washing out oil tanks at sea less advantageous) and then also refuse to let any tanker without that equipment dock in their ports, these actions function similarly to the creation of a club. Most importantly, they give an incentive to states to accept whatever international standards are adopted, since they cannot participate in the benefits unless they do. Also interesting in these examples is the fact that the club gives an incentive to actors who would not inherently gain from the environmental benefit. The owners of oil tankers may not be harmed by oil pollution at sea and therefore gain little advantage from measures to protect the marine environment. But they, and the states on whose behalf they operate, will gain from being able to dock at whatever ports they would like, and joining the 'club' may enable them to do that.

Although rarely discussed as such, there are other elements of environmental cooperation that may have the characteristics of club goods. Environmental monitoring, research, gathering of information, and technological innovation are things that could be proprietary information among states that have joined an agreement. While there has been little evaluation of club goods in the context of environmental cooperation, they may be able to play a role in enticing participation in international agreements.

Lessons

To some extent the limited use of this mechanism within environmental agreements comes from the nature of most environmental issues. Participation in international cooperative efforts may be difficult precisely because most environmental benefits pursued by treaties are those that cannot be denied to non-participants. For those where creation of a club is possible, it has shown promise. Mitchell (1994) points out that not only have states joined the Memorandum of Understanding, but they undertake actions that are much more difficult and intrusive than those required by similar agreements, because they gain a direct advantage from being in the group that does so.

Equipment regulations within efforts to prevent oil pollution provide an example where a common good (protecting the oceans from oil) can nevertheless be sought through creation of a club (those states that use appropriate equipment). At the margin it becomes increasingly difficult to distinguish among threats, side payments, and the creations of club goods, but it is clear that the use of all these instruments, individually or together, have some impact in bringing states into environmental agreements.

CONCLUSION

Although this chapter examines the implications of four different types of mechanisms for encouraging participation in international environmental agreements, it should be acknowledged that the task is analytically difficult. It is not easy to disentangle the membership aspects of these mechanisms; all are used for things other than encouraging participation with treaties. There are reasons to threaten sanctions, give aid, differentiate across states in determining obligations, or, to a lesser extent, create a club good, that go beyond inducing membership. The issue of what it means for a state to join a treaty is also broader than it might initially seem: it may not be sufficient for a state to sign or ratify a treaty if it then uses a treaty's own mechanisms to opt out of the substantive obligations under the treaty. This chapter considers that persuasion of those states to take on full obligations is an aspect of encouraging participation, but realistically it is a different phenomenon than convincing a state to sign and ratify a treaty. Once a state has signed and ratified a treaty, the processes set up within that agreement are responsible for encouraging a state to fulfil its obligations.

It is also not simple to draw conclusions across mechanisms. There is no real way to do a controlled study that would examine which mechanisms work best under which circumstances, since they are not all tried in analogous situations, or even independently. Some agreements, like the Montreal Protocol, make use of all of the mechanisms examined here to some extent; others use only one or two, but in vastly different situations.

It is nevertheless possible to make some cross-cutting observations about the different types of efforts to encourage states to join international environmental agreements. The degree to which these measures are used unilaterally is predictable. Differential obligations are inherently a part of the treaty process, as is the creation of club goods, and aid is most likely used, and most often successful, within the framework of a treaty as well. Sanctions, on the other hand, are rarely used by a treaty organisation, and when they are, the details of imposition are often left up to the individual member states. If having an institutional structure through which these mechanisms are implemented is important, sanctions are likely to be less useful over the longer term than some of the other instruments.

The degree of success is harder to determine. On the one hand, sanctions are far more successful at changing state behaviour within the context of environmental politics than in most other issues in which they are used. The states against whom sanctions are threatened are often those most resistant to joining the treaties in question, and threats of economic restrictions (either inside or outside the treaty) can have a remarkable effect. Aid is certainly helpful not only in persuading states to join but in getting them to implement their obligations under the treaty. Differential obligations are certainly what brought a number of developed countries into the Kyoto protocol, since before the level of differentiation there was no agreement. It may be that differential obligations will become a pre-requisite for developed country participation, in the way that aid has for developing countries. Club goods cannot be provided in all issues, but where possible, they seem to be highly effective at motivating participation.

Because the contexts are different, it is difficult to compare whether one approach is more effective than another. The Montreal Protocol case gives us some evidence, since both differential obligations and trade restrictions included in the original protocol brought a few states in, but the bulk stayed out until a funding mechanism was specified. That would suggest that it is funding that is most successful at bringing states into an agreement. That conclusion makes some sense. With a mechanism like the Multilateral Fund, recipient states are responsible for almost none of the costs of meeting their obligations under the agreement; of course they would prefer that to a system in which they suffer greater costs if they do not join. Likewise, differential obligations still require some action. If states are not interested in undertaking action, it is not clear that giving them ten extra years before they have to meet obligations will change their minds.

All of these mechanisms have implications for the longer run. Sanctions may be less successful overall if states feel (as they do in the whaling case) that they have been bullied into accepting something they do not want. Aid that can help states join and implement treaties is nevertheless becoming more costly as it is now seen as mandatory, and environmental problems cost more to mitigate. Differential obligations may lead to increasing politicisation and complexity of negotiations. Club goods may lead to increased polarisation.

But bringing states into an environmental treaty brings a number of processes into play. It opens up the possibility that they can be persuaded of the importance and possibility of taking action by information about the environmental problem and its solutions, it allows them to participate in the decision-making process that will determine how regulations will be made, and it brings them into the collective plans for implementation. For these reasons, the problems encountered in the course of encouraging states to join international environmental agreements may be outweighed by the advantages of getting them involved.

REFERENCES

R. Abramson, 'U.S. Will Cut Off Trade in Wildlife with Thailand', *Los Angeles Times* 17 July (1991) A9.

Basel Convention on the Control of Transboundary Movements of Hazardous Wastes and Their Disposal (1989).

R. Benedick, *Ozone Diplomacy: New Directions in Safeguarding the Planet* (Cambridge, MA: Harvard University Press, 1991).

T. Bernauer, 'Protecting the Rhine River Against Chloride Pollution', in R. O. Keohane and M. A. Levy (eds), *Institutions for Environmental Aid* (Cambridge, MA: The MIT Press, 1996) pp. 201-232.

J. Bhagwati, 'Aggressive Unilateralism: An Overview', in J. Bhagwati and H. T. Patrick (eds), *Aggressive Unilateralism* (Ann Arbor: The University of Michigan Press, 1990).

D. Brack, *International Trade and the Montreal Protocol* (London: Earthscan Publications Ltd., for the Royal Institute of International Affairs, 1996).

D. Brack, 'The Use of Trade Measures in the Montreal Protocol', paper presented at the Montreal Protocol 10th Anniversary Colloquium, Montreal, 13 September 1997.

P. Brown, 'Playing Football with the Whales', *The Guardian,* 1 May 1993, 26 (Lexis/Nexis).

J.M. Buchanan, 'An Economic Theory of Clubs', *Economica* 32 (1965), 1-14.

C.J. Carr, 'Recent Developments in Compliance and Enforcement for International Fisheries', *Ecology Law Quarterly* 24 (1997) pp. 856-859.

S. Charnovitz, 'Encouraging Environmental Cooperation through Trade Measures: The Pelly Amendment and the GATT', paper prepared for delivery at the Institute on Global Conflict and Cooperation, no date.

S. Charnovitz, 'A Taxonomy of Environmental Trade Measure', *Georgetown International Environmental Law Review* 6(1) (1993) pp. 1-46.

CITES, 'Resolution of the Conference of the Parties: Trade in African Elephant Ivory', (12-24 July, 1987).

Convention for the Prohibition of Fishing with Long Driftnets in the South Pacific, 1989.

Convention for the Protection of the World Cultural and Natural Heritage, 1972.

Convention on International Trade in Endangered Species of Wild Fauna and Flora, 1973.

R. Cornes and T. Sander, *The Theory of Externalities, Public Goods and Club Goods* (Cambridge: Cambridge University Press, 1996).

C. Dasgupta, 'The Climate Change Negotiations', in I. Mintzer and J.A. Leonard (eds), *Negotiating Climate Change* (Cambridge: Cambridge University Press, 1994), pp. 129-148.

E.R. DeSombre, 'Baptists and Bootleggers for the Environment: the Origins of United States Unilateral Sanctions', *Journal of Environment and Development* 4(1) (1995) pp. 53-75.

E.R. DeSombre, *Domestic Sources of International Environmental Policy: Industry, Environmentalists, and U. S. Power* (Cambridge, MA: The MIT Press, 2000).

E.R. DeSombre and J. Kaufmann, 'The Montreal Protocol Multilateral Fund: Partial Success Story', in R. O. Keohane and M. A. Levy, *Institutions for Environmental Aid* (Cambridge, MA: The MIT Press, 1996), pp. 89-126.

E.R. DeSombre, 'Distorting Global Governance: Membership, Voting, and the IWC', in R. Friedheim (ed.), *Toward a Sustainable Whaling Regime* (Cambridge, MA: The MIT Press, forthcoming).

M.P. Doxey, *International Sanctions in Contemporary Perspective* (New York: St. Martins Press, 1987).

L. Elliott, *The Global Politics of the Environment* (New York: New York University Press, 1998).

D. Fairman and M. Ross, 'Old Fads, New Lessons: Learning from Economic Development Assistance', in Keohane and Levy (eds), *Institutions for Environmental Aid: Pitfalls and Promise* (Cambridge, MA and London: The MIT Press, 1996), pp. 29-51.

M. Fineman, 'Dominica's Support of Whaling is No Fluke', *Los Angeles Times* 9 December 1997, p. A1.

H. French, *Partnership for the Planet: An Environmental Agenda for the United Nations* (Washington DC: Worldwatch Institute, 1995).

J. Galtung, 'On the Effects of International Economic Sanctions: With Examples from the Case of Rhodesia', *World Politics* 19 (1967) pp. 378-416.

L. Grant, 'Panama to Clamp Down on Tuna Fish Violators', *Reuters Financial Service* 2 December 1996 (Lexis/Nexis).

G.C. Hufbauer, J.J. Schott, and K.A. Elliott, *Economic Sanctions Reconsidered: History and Current Policy*, 2nd edn (Washington D.C.: Institute for International Economics, 1990).

International Tropical Timber Agreement, 1983.

'Japan to Ban Tuna Imports from Panama, Honduras, Belize', *AP Worldstream* 3 December 1996 (Lexis/Nexis).

K. Knorr and F.N. Trager (eds), *Economic Issues and National Security* (Lawrence KS: Regents Press of Kansas, 1977).

S. Krasner, 'Domestic Constraints on International Economic Leverage', in K. Knorr and F. N. Trager, (eds), *Economic Issues and National Security* (Lawrence KS: Regents Press of Kansas, 1977).

Kyoto Protocol to the United Nations Framework Convention on Climate Change, 1997.

J. Lancaster, 'Endangered Sea Turtle Seen Jeopardized by Japan', *Washington Post* (19 January 1991) A3.

M.P. Malloy, *Economic Sanctions and U.S. Trade* (Boston: Little, Brown, and Company, 1992).

R.B. Mitchell, 'Regime Design Matters: Intentional Oil Pollution and Treaty Compliance', *International Organization* 48(3) pp. 425-458.

Montreal Protocol on Substances that Deplete the Ozone Layer, 1987.

'Panama Denies Violating Tuna Conservation Efforts', *Reuters World Service* 20 November 1996 (Lexis/Nexis).

P.H. Sand, *Lessons Learned in Global Environmental Governance* (Washington D.C.: World Resources Institute, June 1990).

B.N. Schiff, *International Nuclear Technology Transfer* (London and Canberra: Rowman and Allanheld, 1984).

L. Spences, with J. Bollwerk, and R.C. Morais, 'The No So Peaceful World of Greenpeace', *Forbes*, 11 November 1991, 174 (Lexis/Nexis).

K. Stier, 'Thailand Rushes to Pass Legislation on Wildlife Traffic', *Journal of Commerce* 16 July 1991, 6A.

M. Tighe, 'Tuna Commission Approves Sanctions Against Three Countries', *AP Worldstream* 29 November 1996 (Lexis/Nexis).

United Nations Convention to Combat Desertification in Those Countries Experiencing Serious Drought and/or Desertification, Particularly in Africa, 1994.

United Nations Convention on Biological Diversity, 1992.

United Nations Framework Convention on Climate Change, 1992.

United States Code of Federal Regulations 50, 281.2.

United States Federal Register 40 (1975) 32832.

United States Federal Register 48 (1983) 48160.

United States Federal Register 51 (1986) 34159, 36864, 47064.

United States Federal Register 62 (1997) 23497, 44422-3.

United States Public Law 95-376.

UPI Wire Report, 1994.

Part V: Values, Identity and Knowledge

8 Ecocentric Identity and Politics

Ho-Won Jeong and Charlotte Bretherton

In the world which is an intrinsically dynamic and interconnected web, all organisms are constituted by those very holistic ecological relationships. The notion of an ecosystem is represented in the interactive relationships of different forms of animate and inanimate beings. It is essential for the planet's future survival that nature and society have to be dialectically inter-linked and complementing each other. The root of the modern ecological crisis is attributed to disconnectedness of humans with the larger whole. The potential for the development and politicisation of ecocentric identities lies in the way the scale and scope of the contemporary environmental crisis demands a transformative project. The conception that the whole must take precedence over the parts is consistent with transformation in our mode of social and political life.

The key to development of an ecocentric identity becomes the ability to reconceptualise one's own place and the willingness to recognise the identity of human beings as a species within the ecosystem of which we are a part. Different theoretical expressions and approaches exist in understanding the process to extend our sense of self to nature. In this chapter, various sources of ecocentric identification are compared in terms of the extent to which identity politics does not exclude others but nevertheless demands high levels of commitment. The questions of regional, cultural, gender and other social differences are inevitably blended in the discussion of identity. Overall, we are looking at the dynamics of identity construction involving diverse values as well as political and social structures.

ECOLOGICAL IDENTIFICATION AND THE SELF

Though nature is an object of identification, ecological identity refers to a person's connection to the earth, perception of the ecosystem, and direct experience of nature (Thomashow, 1995, p. 3). Given that conceptualisation of non-humans is predicated by reference to the human domain, ecological identity is thus linked to the ways human consciousness is shaped by concerns for the environment. The conventional notion of self can be creatively expanded with a wider construction of identity. Ecocentric identity extends beyond a sense of one's personal self to embrace all beings and ecological processes. Identification with nature is made possible by the transcendence of self (Zimmerman, 1994, p. 293).

In the formation of ecocentric identity, there is neither requirement for alien others, nor can there be claims of superiority over others. More specifically, compared with ethnic and other social identities derived from negative identification with external others, an ecocentric identity does not significantly rely, if at all, upon reference to an external other. In considering that non-humans are recognised as having their own modes of being, the intrinsic existence of non-human nature should not depend on acts of human valuation.

In opposition to reductionism, ecocentrism perceives the world as the body of an expanded self. 'Each ecosystem is a small universe itself in which the interactions of its various species populations comprise an intricately woven network of cause-effect relations' (Taylor, 1998, p. 77). In overcoming human/nature dualism by a process of wider identification, human beings are not merely considered as members of various social groupings but as components of a fundamentally inter-linked and interdependent 'web of nature' (Merchant, 1992, p.86). Since the needs of both the planet and the person have become one with perceptual boundaries breaking down, one's own fate is intimately bound up with that of other species on the planet. In biocentric views which stress direct links between humans and the ecological world, the role of humanity, as an integral part of a much larger whole, is not defined as separate from and in opposition to nature. There are 'no absolute dividing lines between the living and the nonliving, the animate and the inanimate, or the human and the nonhuman' (Eckersley, 1992, p. 49).

Internalisation of ecological understanding based on identification with the biosphere is characterised by the process of discovering a new level of meaning associated with self-actualisation for all beings. With an overall scheme of orientation toward inclusiveness, 'one comes to feel a sense of commonality with all other entities', connected to a single unfolding reality (Fox, 1995, p. 257). The self is thus seen as 'a developing whole, a relative unity-in-diversity, a whole in constant process of self-transformation and self-transcendence' (Clark, 1998, p. 425). Self-realisation to be achieved through extending awareness and receptivity with nature is a transformative process to heal ourselves in the world.

Ecocentric identity does not entail a loss or negation of the self with the dissolution of particular beings but an opportunity to recover one's true nature. In traditional views of identity, this holistic relationship between the self and the cosmos may appear to imply loss of individuality, autonomy and, in particular, self esteem. In practice, however, seeing the self as part of the cosmos would entail widening the sense of self - and hence an expansion in the scope of identity. 'Personal self-realisation is not incomprehensible apart from one's dialectical interaction with other persons, with the community, and with the larger natural world' (Clark, 1998, p. 425). This would be similar to, but much greater than, that experienced through an exclusive identification with a people or nation. Thus, the apparent eclipsing or immersion of the self within the greater whole by taking one's place in a particular bioregional ecosystem or the large biosphere is not experienced as a loss.

Search for a large self involves a constant mediation between one's individual self and the larger whole, as self-actualisation does not create artificial boundaries. The realisation that 'life is fundamentally one' provides a cosmological basis for identification (Naess, 1989, p. 166). An ecological self can be considered not as an entity, or a thing, but as 'an opening to discovering what some call the Absolute' (Devall, 1995, p. 104). Humans and nature are not divided as subject and object, invoking the notion of contingency and participation. Our experience of the world is enriched by the recognition of interconnectedness, interdependence, and diversity of all phenomena. The notion of autonomy is applied to a broader and more encompassing pattern of layered interrelationships in the biotic community, with each layer's role being defined in concentric circles (Fox, 1995, p. 262).

INTUITIVE AND COGNITIVE DIMENSIONS

Development of ecocentric identity requires a new level of intuitive and cognitive understanding of ecology. Psychological maturity is associated with the movement from an atomistic, ego-centric sense of self toward an ecocentric, transpersonal sense of self. With the perceptions of nature being brought to the forefront of awareness, the natural is not conceptually divorced from the human. The affective perceptions of ecological relationships reflect a high level of knowledge and action.

The enlargement of self allows the consideration of a wider set of concerns, and division is healed through a unifying process. Compassion and empathetic feelings toward the natural world flow from the recognition of an individual's way of being in the world. The impetus behind biocentric attitudes is simple, yet enduring, powerful feelings for love of nature with the celebration of another's existence (Thiele, 1999, p. 168).

> ...this love of Nature is no pale intellectual shadow of love, but the real thing...This loving of the world is a blissful state that warms and animates everything around us...It bursts the bars of the personal heart, and vastly expands our sense of self (Mathews, 1991, p.150).

The underlying ideas behind ecocentrism are essentially romantic, reflecting the belief that 'nature and humanity belong in an organic relationship best understood and developed through feeling and insight' (Dryzek, 1997, p.155). Thus ecocentrism, in its various guises, expresses both a preference for emotion over reason and an essentialist view of nature. It is, in consequence, at odds with rationalist/modernist and, indeed, post-modern conceptions of identities as relatively fluid and processual.

The concerns with the preservation of biological species and their support systems constitute part of our basic nature (Southwick,

1996, p. 256). The perception of the expansive circles of interconnectedness leads to a critical reflection and introspection of ecosystem decline. In considering that it is difficult to imagine the world without trees, flowers, birds and other animals, contact with nature is almost universal needs.

Ecocentrism is deeply satisfactory to psychological needs since experiences of commonality with other entities through personal involvement offer potentially great emotional rewards. By ending the fundamental alienation of human beings from their true selves and promoting 'an ineffable sense of at-homeness in Nature, and a disposition to live in harmony with it', an eco-identity brings to its bearer freedom from the 'tyranny of personal desires' and the opportunity to cultivate 'peace and joy' (Mathews, 1991, p.150).

ETHICAL DIMENSIONS

The links to the cosmos can also be developed by the morality based on universal principles of care to take responsibility for others. Special relationships, or empathy, with particular aspects of nature provide the depth of interest that is not otherwise possible. Introduction of ecological projects, both spiritual and material, can depend on an ethic of care for the natural world associated with affective identification. Responsibilities for the ecological world connected to the self can provide an important basis for acquiring a wider, more generalised identity (Zimmerman, 1994, p. 288).

By encountering nature simply as a collection of neutral things, we deny it 'any sense of the character of worldhood' (Evernden, 1993, p. 66). Ethical and spiritual identification with nature is made through non-utilitarian values such as love of nature, in that the appreciation of the material usefulness of nature alone generally does not produce a sense of attachment. Ecological values, aesthetic and philosophical convictions are different from a one-dimensional utilitarian end towards the satisfaction of rational self-interest. The expression of ethical norms and values is 'motivated by strong feelings' as well as 'having a clear cognitive function' (Naess, 1989, p. 64).

To repudiate the notions of human superiority necessitates a fundamental reassessment of the essence of the self, revealing the internal and external referents of one's own identity. Forming a wider sense of identity permits people to care for the ecosystems instead of treating them as mere commodities. It challenges the instrumental treatment of the other based on sharply defined ego boundaries which stress the separation of humans from other species.

Moral dimensions of ecological projects require sympathy with and sensitivity to the situation and fate of particular others. People, as moral agents, are expected to embrace the obligation to respect both living and non-living objects on the planet. An environmental ethic does not derive from the technological vision of the world but naturally emerges from love for what humans identify themselves as part of (Evernden, 1993, p. 69; Devall, 1995, p. 104). The fundamental sense of interconnectedness between all life forms implied by an ecocentric identity rejects modernist conceptions that consider the object in terms of self-interest.

The expanded self has to be based on a critique of modern egoism rather than its extension. The denial of human superiority is made by recognising each individual organism as 'a teleological center of life, pursuing its own good in its own way' (Taylor, 1998, p. 74). The basis for wider and deeper identity can be created by unbridling egoism of the particular and conceives of all living things as entities possessing inherent worth.[1] The essential core of deep ecology therefore criticises the self rooted in rational egoism that derives from the anthropocentric, materialist assumptions of western Enlightenment thought. Since the ideologies of modernity, in particular liberal capitalism, destroyed the perception of the earth as a living being, an industrial society tends to overlook the fact that humans are part of nature.

Ecocentrism guided by the norms of self-realisation for all beings emphasises a shift away from anthropocentric humanism that puts nature into a realm of instrumentality (Zimmerman, 1994, p. 2). An ethical position of ecocentrism lies in treating all of the multilayered parts of the biotic community as valuable for their own sake (Eckersley, 1992, p. 28). Epistemological shift for biospheric egalitarianism illustrates the fact that while humans have a unique capacity to abuse and exploit other species and, ultimately, to destroy the planet, they do not have a greater intrinsic value than any others.

The discovery of an ecological self leads to its defence and interaction with it. Reordering human relations with nature, in all its facets, follows the reflection of the true position of humans. With the disappearance of distinctive differentiation between higher and lower life forms in nature, particular entities should not impose their will unduly upon others (Fox, 1995, p. 256). Humans have to learn how to live in harmony with non-human species and the ecosystems of which they are all a part.

A SOCIAL CONTEXT

Ecological politics cannot ignore the implication of ecocentric identity for the transformation of social relations and its ultimate impact on nature. Beyond spiritual commitment and experience, political questions are involved in the determination of our links to nature. As discourse on nature is placed into a broad social context, ecological crisis is attributed to the social order itself. The deadly destruction of the natural world is rooted in the creation of human hierarchies.

Whereas deep ecology stresses nature's equal right to existence, social ecology focuses on a political community of humans. The persistence of power structures cannot be separated from views concerning nature. The natural world has been pillaged for prestige, profit and control in an inegalitarian system that allows the elite to subjugate other people. The structuring of the natural world into a vertical chain of being originates from the establishment of authoritarian social relations. With the rise of human domination, 'the seeds are planted for a belief that nature not only exists as a world apart, but that it is hierarchically organised and can be dominated' (Bookchin, 1993, p. 365).

The understanding of ecological identity not only serves as but also depends on the critique of unjust social order. Social ecologists see the structures of social inequality as both prior to, and a source of, domination over nature. Thus the question of our proper place in the rest of nature is not distinguished from the issue of an appropriate type of social and political arrangements. Hierarchy, as a social construction of a stabilising or ordering principle, does not have a natural basis (Bookchin, 1982). Instrumental values, responsible for the exploitation of nature, are associated with the control of human behaviour that limits freedom and prevents self-realisation.

In considering that power relations and cultural norms of a contemporary society legitimise a dangerously dysfunctional relationship between human collectivities and the ecosystems, the existing social structure needs to be transformed to eliminate the oppression of both nature and humans. The acceptance of the equal worth of all life forms as well as the essential inter-relatedness and interdependence of the ecosystems would inevitably result from the rejection of all forms of exploitation. In social ecology, ecological issues are approached through campaigns against the structures of social injustice that are deeply embedded in hierarchical relations.

The well-being of the natural world is bound up with the creation of egalitarian human relations based on democratic political and socio-economic arrangements (Zimmerman, 1994, p. 2). Social malaise and the traditional distance from the natural world can be overcome by decentring the human world. Ecological identity can be regarded as truly emancipatory through its social links - it would end both alienation and domination. At the same time, the bifurcation between the social and natural realms would dissipate by transposing the non-hierarchical character of natural ecosystems to society.

An ecological identity is compatible with social organisations that encourage symbiosis, diversity, interconnectedness, stability and flexibility. For transformative politics, considerable efforts can be made to dismantle unequal structures by bringing about the changes in human values and behaviour. Ecological sustainability is attainable in grassroots democracy with decentralisation and autonomy (Doyle and McEachern, 1998, p. 39). More specifically, ecological concerns are more satisfactorily met in small scale, egalitarian, anarchistic societies.

FEMININE IDENTIFICATION WITH NATURE

Instead of taking universalistic positions, ecofeminism examines 'those domains of human experience that have been relegated to women: namely the personal, the emotional, the sexual' in seeking to enlarge our understanding of nature (Hallen, 1995, p. 203). The form of identification is gender specific with the ecological self associated with femininity. The alien other of masculine identity is negatively defined against feminine identity associated with the realm of the physical. The structure of power infused with masculine norms and values brings a certain equivalence to the relationship between women and the natural world. The readiness for this special affinity arises from an analogy between women's position of being oppressed and the destruction of nature. The work of nature is unappreciated by human existence in the way women are overshadowed. The feminine character of nature is, therefore, contrasted with the masculine values of dominant Western cultures.

Gendered structures of power and inequality underlie and legitimise disjunctures between human organisations and ecosystemic needs.[2] Humans are not equally alienated from, and exploitative of, the natural world. The logic behind the inferiorisation of non-human nature and the oppression of women has been justified by patriarchy which serves as a symbolic, historical source of oppression (Warren, 1998, p. 328). In masculine values and their instrumentalism, women and nature are regarded as subjugated other. Through a special convergence of interests between feminism and ecology, the devaluation of nature is interpreted as a product of masculine consciousness (Eckersley, 1992, p. 64).

The patriarchal aspect of modernity has been promoted by a totalising attitude that seeks to erase difference in order to attain a problematic unity. The domination of an ecological system 'as feminine' is constructed on capitalist patriarchal culture of modernist discourse that treats nature merely as material for man's appropriation.[3] As nature is tamed and subjugated to masculine will, there is a necessary connection between patriarchy and anthropocentrism. In the logic of these dualist systems of thought, identification of men with the 'human' and the realm of the mental permits transcendence - superiority over, freedom from, control of nature (French, 1985, p.xvi). The rationalist conception of the modern self formed through masculine values is grounded in anthropocentric, human self-interest.

Empowerment for women is based on a critique of the twin oppression of women and non-human nature. Feminist interest in emancipation from the status of otherness through the elimination of patriarchy translates into efforts to free nature (Zimmerman, 1994, p. 2). At the same time, an ethic of care for the environment reflects women's experience of vulnerability and interdependence. Many Third World women's struggle to protect indigenous forests against the introduction of commercial agricultural production is related to not only their economic subsistence but also cultural survival.

The core of ecological identity can be supported and constituted by the feminine notion of cooperation, mutual support and nurturance as opposed to masculine competitive values. The earth and women, equated with maternalist values, are revered as the nurturers of life against hierarchy and destruction. The recovery of a spiritual relationship with the earth, as mother, is essential to an ecocentric worldview, and indeed to the future of the planet, for 'one does not readily slay a mother, dig into her entrails for gold or mutilate her body' (Merchant, 1982, p.3). In some non-western cultures, maternalist perceptions of the earth as the nurturers of life continue to influence the behaviour of humans. [4]

BIOREGIONAL IDENTITY AND CULTURE

In bioregional perspectives, an ecological identity is based upon a specific sense of place with the expression of its concerns through very specific and local responsibilities of care. Cultural norms are a central aspect of human experience that mediates the relationship between social and natural systems (McGinnis, 1999, p.5). In considering that ecological systems can be understood in terms of characteristics of specific regions and related human responses, a bioregional identity derives from belonging to and forming part of the local ecosystem. [5] Ecological integrity emerges in each region with its own unique identity.

Bioregional identity is formed by the interaction between geographical characteristics and cultural value systems. [6] Increased differentiation from other regions would be necessary for cultural homogenisation within communities that inhabit a particular bioregion. With the understandings and practices of human collectivities tied to the boundaries of a local ecosystem, the content of ecological identity is contingent upon people's unique connections to a particular place and its natural world.

Through a learning process, human values and norms will be gradually adapted to the needs of local ecosystems. Ecological identification can be based on highly particularistic meanings attached to 'a place that motivates passion of many modern conservationist and love of many indigenous peoples for their land' (Zimmerman, 1994, p. 297). In particular, cultural adaptation to the ecological needs of local places is well revealed in the thinking and practices of indigenous peoples.[7]

Despite the estrangement of indigenous culture from nature following modernisation and industrial capitalism, the understanding of how to live in, rather than with, nature is not completely lost. Rather it is deeply buried in cultural memory, so that 'Knowledge of place, within us, needs to be uncovered and revered' (McGinnis, 1999, p.9). This is achieved through a process of 'reinhabitation', which involves 'becoming native to a place through becoming aware of the particular ecological relationships that operate within and around it' (Berg & Dasmann quoted in Eckersley, 1992, p.167). It also entails 'a kind of ecological citizenship', encouraging individuals to 'learn to become respectful' of 'an ecological place, rather than transforming the place to suit themselves' (Dryzek, 1997, p.160).

In ecological thinking, interdependence provides a fundamental basis for ecological diversity with variations in genes, niches and adaptive traits facilitating species survival. Since cooperation between bioregions is essential to safeguarding larger ecosystems, the locally based ethic of care for the bioregion should not undermine the characteristics of other communities. On the other hand, the experiment in the local community can be applied outside of that locality, and the experience can be compatible with the maintenance of a large ecological system. Regional ecological sustainability does not necessarily have to involve competitive needs with other regions due to the fact that qualitative development can be achieved through differentiation.

Bioregionalism differs from deep ecology in ways that enable it to 'straddle green romanticism and green rationalism' (Dryzek, 1997, p.160). Its local focus could prove important for social learning. On the other hand, bioregional perspectives can be criticised for a lack of a pragmatic adaptation to present circumstances, and have been pointed out as naive 'anarchic primitivism' that, beyond a way of thinking, is not reflected in an actual way of living (Aberley, 1999).

In modern societies, community cohesiveness, the preservation of the local bioregion and the protection of indigenous life forms have been diluted by the influx of new residents and the import of non-native plants or animals. While the framework for experiments with 'living in nature' has to derive from celebrating an important aspect of human experience, our recent sentiment towards the gradual demise of wildlife is related to increasing rootlessness represented by experiences of greater mobility and the dissolution of local communities (Cooper, 1993).

POLITICISATION OF ECOCENTRIC IDENTITY

Identity politics involves the prioritisation of one particular facet of identity over others, in a manner that influences political choices and, potentially, provides a basis for social action. The prescriptions of ecocentric identity are consistent with notions of identity politics, in which a dominant identification is appealed to and acted upon (Bretherton, 2001). The impact of an ecocentric approach is profound given its role in translation of environmental anguish into a radical transformation of human identity. In ecocentrism, the solution to ecological crisis is sought in broad identification with an appeal based on love, inescapable duty and potential sacrifice. The impulse to defend the existential rights of wilderness stems from a great sensitivity to the needs of various life forms as well as individual communities.

Various interpretations of humans' position within the broader ecosystem and other sources of identification have different implications for transformative politics. Opportunities for structural transformation do not arise from the contingent and shifting base of identification, whereas essentialist treatments of identity are conducive to commitment, even self-sacrifice. The relationship of the self to others is represented and expressed in a non-conformist manner by powerful claims about identity and community contained in green culture (Deudney, 1993, p. 295). A change in individual action based on reorientation toward a new relationship with the natural world would ultimately be demanded by the politicisation of identity.

Ecofeminists see the expanded self as the completion of a process of masculine universalisation, moral abstraction and disconnection. The emphasis on 'wider identification' is interpreted as the self-expansion of the modern masculine ego. In deep ecologist views, on the other hand, the manifestations of general anthropocentric humanism do not necessarily disappear with liberation from oppressive gender or class relations of patriarchy (Zimmerman, 1994, p. 9, 10). Moreover, a claim for a more encompassing ecocentric identity would not easily derive from maternalist essentialism that has exclusive identification of women with nature.

Homogeneous ecological identity is difficult to achieve due to the social divisions of race, class and gender.[8] The strength and endurance of existing social institutions reduce the impact of the political appeal of universalistic approaches to ecological identity on the public. The principal focus of an emancipatory ecological praxis assumes parallels in challenges to the logic or symbolic structure of different kinds of domination. It directly questions deeply entrenched cultural/religious understandings concerning what it means to be a human being - a member of a superior species at the pinnacle of the evolutionary process and answerable only to other humans or, perhaps, the gods. Implicit, here, is our view that such norms and values would also be conducive to social justice.[9]

The boundary between the natural world and society remains a contested interface with the impact of culture on the way people construct representations of their social and physical environment. Due to differences in interpretation of culture/nature relationships both across and within societies, ecological thought defies categorisation into traditional schools. Assessing the positions of humans in nature is not able to escape infinite variety. Identification with cosmopolitan or universal values does not easily supplant more particularistic identities such as affective ties of kinship, although kinship based attachments can be inferiorised by universalisation of the superiority of reason. A relative lack of an emphasis placed on personally based identification leads to ignorance of particular connections.

Social and personal identity strengthens feelings of an attachment to particular areas of land, yielding ties as special and powerful as emotional attachments to kin. Whereas bioregionalist concerns with the development of a sense of rootedness are considered to be 'biotic, not merely ethnic' (Morris Bergman, quoted in Eckersley, 1992, p.168), the potential to politicise a sense of place along ethnic lines remains strong. The bioregionalist appeal to a local identity can be easily supported by the primordialist emphasis upon affective identification with the known and exclusion of the unknown and unfamiliar.

By being marked by a sense of ownership as well as a sense of belonging, an environmental identity subsumed within primordialism encourages the extension or enlargement of individual identity to a nation (Grosby, 1994, p.165). The rewards for the individuals offered by ethno-national or religious identification are significant because they include not only a sense of membership in a wider community but also a sense of continuity with the past, and with the future, which serves to assuage the knowledge of human mortality (Anderson, 1991; Balakrishnan, 1995).

The close association between ethnic identification and home place would overshadow the formation of broad ecological identity. In fact, the development of a local identity could, in some circumstances, be diverted into an ethnic politics of exclusion. Ethnic closure against outsiders may happen with the cultural homogenisation of communities through adaptation to local ecosystems. The politicisation of regional identities along ethnic lines would engender conflicts of interest between bioregions.

Narrowly based ecological identity thus does not prove to be a satisfactory solution to pressing transborder problems, since it is alien to a broader ethic for complex and interrelated ecosystems extending beyond one's own borders. Personally based identification can slip so easily into parochial attachment and proprietorship. Territorial governance is inadequately and poorly equipped to foster borderless ecocentric identity. The prospects for the sustainable environment are not bright if a nation-state, detached from the planetary whole, remains a dominant form of governance with its power for struggle to control land and other resources.

Despite its variants, radical ecologism is opposed to managerial environmentalism whose goal is limited merely to seeking the preservation and efficient use of resources. It questions the organising principles and value systems that sustain modern political and economic life. Beyond challenging the virtues of modern rationality that justifies destructive treatment of nature, most importantly, ecocentric identity is geared toward restoring the complexities of the ecological web of life as a politically salient goal.

ECOLOGICAL POLITICS

Human individuals' sense of who they are and where they belong is involved in prioritising ecologically functional ways of being and constructing societies whose shared norms and values contribute to a harmonious relationship between nature and culture. It can be argued that the identification with a large whole should not be a purely logical process. The framework of ethical universalisation may be seen as being seriously incomplete if it is reduced simply to duty or obligation.

Ecocentrism's challenge to modern orthodoxy has been widely resisted and misrepresented given that it is seen as the idealisation of nature. For many individuals, the radical change of values and of consciousness would undoubtedly be perceived in terms of sacrifice and loss. A search for reconciliation between romantic attachments to nature and rational self-interest would be more acceptable to the public. Humans may be rendered self-conscious about their stewardship by ecological values. The appeals to preserving the wilderness can be made by justifications, ranging from protection of wilderness as a life support system for humans to aesthetic and religious experiences of wilderness.

Ecocentrism stresses, as opposed to an anthropogenic perspective, that a large ecological community should not be treated as serving instrumental value in human decision making. It does not necessarily mean that humans have to live in impoverished conditions, nor does it imply a complete abandonment of cultural heritage (Naess, 1989). The spirituality of place, non-human rights and other emphasis on holism are used for mobilisation of political support for ecological movements. In the post-material values of the 'new middle class', environmental protection produces aesthetically pleasing outcomes or otherwise enhances the quality of human life (Hannigan, 1995).

The political programme advocated by the proponents of ecocentrism stresses primarily the lifestyle of individuals and communities with a focus on 'authenticity', 'simplicity' and 'naturalness' for the self and an example to others. For deep ecologists, 'politics is not about devising strategies to achieve tangible goals; rather it is an arena in which different kinds of experiences can be sought and developed' (Dryzek 1997, p.155). An ecocentric identity and associated ways of living thus become not simply the means, but also the desired end, of deep green politics. From this perspective, the future of the planet is a matter of strong personal commitment. In consequence, the capacity of this 'politics of identity' to extend its influence becomes crucial with education towards expanding the self.

NOTES

1. Questions of intrinsic value are important, and contested, in green political theory/philosophy. They are discussed in some depth by Mathews (1991) and Goodin (1992) amongst others.

2. A number of ecofeminists deal with this issue – see, for example Plumwood (1994). A focus upon gendered structures of political and economic power, and their implications for environmental politics, can also be found outside the ecofeminist tradition. See, for example, Seager (1993) and Bretherton (1998).

3. Historian of science, Caroline Merchant (1982) has documented changing European attitudes towards the natural world, and towards women, from the 16th Century onwards, illustrating the gradual rejection of a holistic worldview in favour of a dualistic system of thought in which (female) nature became divorced from (male) culture.

4. For an interesting case study, in the context of India, see Marglin & Mishra (1993).

5. Since traversing or dissecting existing political boundaries of bioregions is difficult, sets of 'bioregional overlays' have been proposed, for the guidance of local community action. Ecological/cultural boundaries of bioregions might possibly lead to 'alternative political boundaries' with the establishment of new ways of living (Klyza, 1999, p. 92).

6. As in the case of ecofeminism, there is no single, definitive bioregionalism. For a wide ranging discussion of the antecedents and current manifestations of bioregionalism, see McGinnis (1999).

7. Interestingly, a recent study of conservation schemes in Central America (which sought, *inter alia*, to maintain indigenous autonomy) reached the conclusion that the problems encountered 'may temper bioregionalism's exaltation of the indigenous lifestyle' (Ankersen, 1999, p.184).

8. This is graphically illustrated by Lois Gibbs, a spokeswoman for the (also locally based, in the US) environmental justice movement - 'Environmentalists are people who eat yoghurt, while my people drink Budweiser and smoke' (Quoted in Dryzek, 1997, p.178).

9. As Dobson concludes (1998) there is, in the thought of many deep ecologists, a tension between ecological imperatives and social justice - with a clear tendency to prioritise the former.

REFERENCES

D. Aberley, 'Interpreting Bioregionalism: A Story from Many Voices'. In M.V. McGinnis (ed), *Bioregionalism*, (London: Routledge: 1999) pp. 13-42.

B. Anderson, *Imagined Communities: Reflections on the Origins and Spread of Nationalism* (London: Verso, 1991).

T.A. Ankersen, 'Addressing the conservation conundrum in Mesoamerica: A bioregional case study', in M.V. McGinnis (ed), *Bioregionalism* (London: Routledge, 1999) pp. 171-88.

G. Balakrishnan, 'The National Imagination', *New Left Review*, 211 (1995) pp. 56-69.

M. Bookchin, *The Ecology of Freedom: The Emergence and Dissolution of Hierarchy* (Palo Alto, Calif.: Cheshire, 1982).

M. Bookchin, 'What Is Social Ecology?' in M. Zimmerman (ed.), *Environmental Philosophy* (Prentice Hall, Englewood Cliffs: Prentice Hall, 1993) pp. 354-73.

C. Bretherton, 'Gender and Environmental Change: Are Women the Key to Safeguarding the Planet?' In J. Vogler & M. F. Imber (eds), *The Environment and International Relations*, (London: Routledge, 1996) pp. 99-119.

C. Bretherton, 'Global environmental politics: putting gender on the agenda?', *Review of International Studies*, vol. 24, no. 1 (1998) pp. 85-100.

C. Bretherton, 'Ecocentric Identity and Transformatory Politics', *International Journal of Peace Studies*, vol. 6, no. 2 (2001).

J. Clark, 'A Social Ecology', in M. Zimmerman (ed.), *Environmental Philosophy* (Upper Saddle River, 1998) pp. 416-40.

D. Cooper, 'Human Sentiment and the Future of Wildlife', *Environmental Values*, vol. 2, (1993) pp. 335-46.

D. Cooper and J. Palmer, *Just Environments: Intergenerational, International, and Interspecies Issues*, (New York: Routledge, 1995).

B. Devall, 'The Ecological Self', in A. Drengson and I. Yuichi (eds.), *The Deep Ecology Movement* (Berkeley: North Atlantic Books, 1995), pp. 101-123.

A. Dobson, *Justice and the Environment: Conceptions of Environmental Sustainability and Dimensions of Social Justice* (Oxford: Oxford University Press, 1998).

T. Doyle and D. McEachern, *Environment and Politics* (London: Routledge, 1998).

J.S. Dryzek, *The Politics of the Earth* (Oxford: Oxford University Press, 1997).

R. Eckersley, *Environmentalism and Political Theory* (London: UCL, 1992).

N. Entessar, 'The Kurdish mosaic of discord', *Third World Quarterly*, vol. 11, no. 4 (1989) pp. 83-100.

N. Evernden, *The Natural Alien: Humankind and Environment* (Toronto: University of Toronto Press, 1993)

W. Fox, *Toward a Transpersonal Ecology*, (Albany: State University of New York Press, 1995).

M. French, *Beyond Power: On Women, Men and Morals* (London: Sphere Books, 1985).

E. Gellner, *Culture, Identity and Politics* (Cambridge: Cambridge University Press, 1987).

R.E. Goodin, *Green Political Theory* (Cambridge, Polity, 1992).

S. Grosby, 'The Verdict of History: the Inexpungeable Tie of Primordiality', *Ethnic and Racial Studies*, vol. 17, no. 1 (1994) pp. 164-71.

J.A. Hannigan, *Environmental Sociology* (London: Routledge, 1995).

C.M. Klyza, 'Bioregional possibilities in Vermont', in M.V. McGinnis (ed), *Bioregionalism* (London: Routledge, 1999).

M.V. McGinnis, 'A Rehearsal to Bioregionalism', in M.V. McGinnis (ed.), *Bioregionalism* (London: Routledge, 1999) pp. 1-10.

F.A. Marglin and P.C. Mishra, 'Sacred Groves: Regenerating the Body, the Land and the Community', in W. Sachs (ed), *Global Ecology* (London: Zed Books 1993) pp. 197-208.

F. Matthews, *The Ecological Self* (London, Routledge, 1991).

C. Merchant, *The Death of Nature: Women, Ecology and the Scientific Revolution* (London: Wildwood House, 1982).

C. Merchant, *Radical Ecology* (London: Routledge, 1992).

A. Naess, *Ecology, Community and Lifestyle*, translated by D. Rothenberg, (Cambridge: Cambridge University Press, 1989).

V. Plumwood, 'The Ecopolitics debate and the politics of nature'. In K.J. Warren (ed), *Ecofeminism* (London: Routledge, 1994) pp. 42-63.

D.I. Roussopoulos, *Political Ecology: Beyond Environmentalism* (Montreal: Basic Books, 1993).

J. Seager, *Earth Follies: Feminism, politics and the environment* (London: Earthscan, 1993)

P. Taylor, 'The Ethics of Respect for Nature', in M. Zimmerman (ed.), *Environmental Philosophy* (Upper Saddle River, 1998) pp. 71-86.

L.P. Thiele, *Environmentalism for a New Millennium* (New York: Oxford University Press, 1999).

M. Thomashow, *Ecological Identity*, (Cambridge, Mass.: MIT Press, 1995).

K.J. Warren (ed.), *Ecological Feminism*, (London: Routledge, 1994).

A. Tickle & I Welsh, 'Environmental Politics, civil society and post-communism', in A. Tickle & I. Welsh (eds), *Environment and Society in Eastern Europe* (Harlow: Addison, Wesley, Longman, 1998) pp. 156-85.

M.E. Zimmerman, *Contesting Earth's Future* (Berkeley: University of California Press, 1994).

9 Globalised Networks of Knowledge and Practice: Civil Society and Environmental Governance[1]

Ronnie D. Lipschutz

I. INTRODUCTION

There is little doubt that the global environment is under stress, as we observe that climate is changing; soil is eroding; forests are disappearing; species are dying.[2] Who damages the environment? Why do they do it? What is to be done? Who is to do it? About ten years ago, the Public Broadcasting System televised a miniseries, entitled *After the Warming*, in which the answers were provided. With James Burke as writer/commentator/guide, the program provided graphic illustrations of an imagined future in 2050, looking back upon a chaotic past disrupted by global warming. The impacts were, of course, widespread and horrific: How else to illustrate the premise of the show? But the proferred solution was a somewhat hubristic 'Planetary Management Authority' (PMA) run, 'of course' (as Burke put it), by Japan (a prediction that now seems rather hollow).

The PMA – housed in a futuristic building which, even today, looks garish and anachronistic – would consist of a formidable system of computers and sensing devices that, utilising a complex climatic model, could assess the impacts on global climate of human activities all over the planet. As necessary, the PMA would then issue appropriate directives to mitigate or ameliorate the climatic consequences: A global panopticon, in other words; not a

World State but a *World Manager*, with complete authority and power.

Such a centralised management system is quite improbable. It flies in the face not only of logic but also contemporary global politics that are characterised as much by fragmentation within existing polities as by global economic integration among them. The relationship between integration and fragmentation suggests that 'global environmental management' will be even more difficult than has been so far practiced or projected, especially if the number of sovereign or 'semi-sovereign' entities participating in international politics – and subject to transboundary effects – were to increase over the coming decades from the fewer than 200 we now have to many more. This is not the only complication. Many of these entities do not even exercise effective political control over their own juridical territories, much as is presently the case in various 'failed' states, subject to ethnic wars and other types of conflict, around the world (see, e.g., Kaplan, 1996; Barber, 1995). As a result, environmental protection, and the governance it requires, may well require much more substantial and extensive political arrangements of global scope than the international cooperation among states that is today's conventional wisdom (Haas, Keohane & Levy, 1993; Keohane & Levy, 1996).

The answers to the questions I posed at the outset of this chapter are, therefore, not at all obvious. What, after all, do we mean by the term 'global environment'? For that matter, what do we mean by the terms 'global' or 'globalised'? Thinking about these questions – let alone finding answers – requires a reconsideration of their framing and meaning as well as a much better understanding of how environmental action is operationalised in a globalising world. Not only do we need to better understand the *physics* and *biology* of environmental degradation, we must also come to comprehend its *sociology* and *politics* for, although we might bemoan the crudities and imprecision of the global circulation models that warn us of the possibilities of climate change, our understandings of human *social* behavior and change are far cruder (Redclift & Benton, 1994; Lash, Szerszynski & Wynne, 1996; Lipschutz 1996b: ch.2-3). But make no mistake, it is social change that will be needed if the Earth's environment is to be protected and sustained, and how that change is to come about is the critical question for the 21st century.

In this chapter, I offer one view of how such change could come about and argue that it is already under way through what I and others call 'global civil society'. As conventionally understood, civil society includes those political, cultural and social organisations of modern societies that have not been established or mandated by the state or created as part of the institutionalised political system of the state (e.g., political parties), but are nevertheless engaged in a variety of political activities.[3] Globalising the concept extends this arrangement into the transnational arena, where it constitutes a proto-society composed of local, national and global institutions, corporations, and nongovernmental organisations, many of which are engaged in activities related to the global environment.

I begin this chapter with a discussion of the fundamental problems that globalisation poses to the regulatory capabilities and will of nation-states and, in turn, to the global environment. I argue that, inasmuch as national regulation no longer suffices to deal with a broad range of transnational and transboundary issues, and that 'international regimes' are necessary but not sufficient to this task, there is a growing need for more comprehensive regulatory arrangements of global scope but, so far, a lack of appropriate institutional forms. In the second part of the chapter, I discuss several ways in which such arrangements could emerge. The framing of transboundary problems in terms of cooperation among states is an important first step in addressing those problems, but it does not go far enough. I propose that global civil society represents one emerging part of a system of global governance that can include states, but sometimes does not, and describe some of the environmental regulatory activities of 'global civil society', mediated through networks of both knowledge and practice. In the last part of the paper, I offer some ideas about the future of such global governance and how it might be facilitated.

Several initial notes: Throughout this chapter, I use the term *globalisation* to denote both a material and an ideological/cognitive process (see Castells, 1996, 1997, 1998). Globalisation is *material* in the sense that it involves the movement of capital, technology, goods and, to a limited degree, labour to areas with high returns on investment, without regard to the social or political impacts on either the communities and people to which it moves or to those left behind. Globalisation is *ideological* in the sense that such movement is rationalised in the name of 'efficiency, competition and profit'. And, globalisation is *cognitive* in the sense that it fosters social innovation and reorganisation in existing institutions, composed of real, live people, without regard to the consequences for them. In both regards, although globalisation opens numerous political opportunities for social movements and other forms of political organisation and action, a not uncommon result is disruption to existing forms of beliefs, values and behaviors.

Knowledge is defined here as a system of conceptual relationships – both scientific and social – that explains cause and effect and offers the possibility of human intervention and manipulation in order to influence or direct the outcomes of certain processes. *Local knowledge* encompasses such knowledge, as well as the specific and *sui generis* social and cultural elements of bounded social units. *Civil society* includes those political, cultural and social organisations of modern societies that are autonomous of the state, but part of the mutually-constitutive relationship between state and society. *Global civil society* extends this concept into the transnational realm, where it constitutes something along the lines of a 'regime' composed of local, national and global nongovernmental organisations. Finally, *governance* is, in Ernst-Otto Czempiel's (1992, p. 250) words, the 'capacity to get things done without the legal competence to command that they be done.' In this sense, it is a form of *authority* rather than jurisdiction (see Lipschutz, 1996b, 1999b).

II. WHO RULES? WHOSE RULES?

There is not much question that the global demand for environmental protection is growing, even as disputes over the extent and timing of environmental change, and burden-sharing among countries, hobble international efforts to establish effective regulatory arrangements. In a number of arenas, in recent years, efforts in negotiating fora have slowed almost to a standstill, or entirely fallen apart, because of disagreement over basic economic and political propositions. This disagreement is a consequence of two aspects of globalisation. First, in an increasingly industrialised and consumption-oriented world, environmental damages that were once limited to national territories are, increasingly, transboundary and global. Second, in an increasingly globalised economy, paying for the costs of social externalities is seen as a drag on profits and competitiveness.

These difficulties are exacerbated by a phenomenon that James Rosenau (1990) and others have tagged with the somewhat ungainly term 'fragmegration.'[4] On the one hand, globalisation suggests the notion of a single world, integrated via the global economy, in which the sovereign state appears to be losing much of its authority and control over domestic and foreign affairs (Ohmae, 1995; Woodall, 1995; Strange, 1996). This trend seems to point toward a single world market in which economics dominates politics. But, contrary to the expectations of neo-functionalists and others, global economic integration does not seem to be generating a parallel process in the political realm, where we see instead the fragmentation of authority into ever-smaller pieces (Lipschutz, 1999b).

On the one hand, states are decentralising, deregulating and liberalising in order to provide more attractive economic environments for financial capital (for an argument that such deregulation has *not* taken place, see Vogel, 1996). On the other hand, as governments proceed along this path, the safety net provided by the welfare state is being dismantled. That safety net, it should be noted, includes not only guarantees of health and safety, environmental protection, public education, and so on, but also standard sets of rules that 'level the economic playing field' and ensure the sanctity of contracts. The latter two are especially important to capital. Unfortunately, one of the results of such 'leveling' is growing social externalities that capital refuses to pay and which cannot or will not be addressed by nation-states acting alone.

Protection of the environment is one arena in which the externality problem is particularly evident, especially in cases in which activities within one state have impacts on environments within others. The conventional solution for this lacuna is cooperation between or among countries, through 'international regimes' that function to monitor and regulate offending activities (see, e.g. Young, 1994; Haas, Keohane & Levy, 1993; Keohane & Levy, 1996). But while international environmental regimes have shown some successes, especially in controlling ozone-depleting substances, other negotiations have proved much more difficult.

As economic competitiveness and growth have become more important to the domestic politics of industrialised countries, moreover, the costs of dealing with global environmental externalities have also come to seem more onerous. Countries such as the United States are faced with multiple 'constituencies', both domestic and international, that must be satisfied. At home, interest groups apply pressure, both pro and con, to the national government; abroad, various states do the same (Leatherman, Pagnucco & Smith, 1994). Policy makers seek to conclude deals that will minimise the costs they have to pay while maximising the benefits for all. The result is that not much happens.

Not all issue areas are subject to such intractable 'two-level games' (Putnam, 1988); one way to address this particular problem is to narrow the boundaries of political contestation. To a growing degree, nation-states have begun to yield a substantial amount of their domestic regulatory authority to transnational regulatory regimes and organisations, such as the International Monetary Fund and the World Trade Organisation, and to private corporate initiatives such as 'Responsible Care' and ISO 14000. These arrangements are intended to accomplish two goals, both of which exclude popular participation or representation. First, they are meant to reduce the transaction costs to capital associated with a plethora of national regulations that reflect cultural and social differences among countries. Second, and of greater concern here, they are also meant to 'eliminate politics' from certain potentially-conflictual issue areas by shifting regulatory authority from the domestic sphere to the international one, where representative national and sub-national institutions lack power and influence.

The way in which exclusion occurs can be seen, in particular, in campaigns to extend international regulation to include a variety of social matters. Such proposals are often strongly opposed by government authorities, corporate officials and a number of prominent academics, who often argue that supranational rules are acceptable if they involve barter, banking, budget deficits or borrowing, but inappropriate if environmental protection, human rights, labour standards or distributive justice are involved (see Zaelke, Orbuch & Houseman, 1993, especially Bhagwati, 1993). The fact of the matter is that existing international regulatory law does not so much eliminate politics from contentious issue areas as it privileges the political desires and goals of transnational capital and corporations. By limiting debates to small groups of national representatives and company executives, and 'letting the free market do it', most international regulatory arrangements have deliberately been made both opaque and non-democratic as well as quite limited in scope. With a few exceptions, where transnational social regulations have been promulgated, the direction of harmonisation has been more in the direction of the 'lowest common denominator'.

This trend toward international regulation has contradictory consequences in the case of the global environment. Environmental negotiations and regulatory efforts have been far more open and transparent than has been the case for other issue areas, especially trade and finance (although even this is changing), and the presence of nongovernmental organisations, corporations and others is deemed essential, if only to legitimate international agreements. At the same time, however, such participation is often severely circumscribed. At the end of the day, the structure of an international environmental regime may actually serve to institutionalise practices that caused the problem in the first place, rather than controlling or eliminating the offending practices.

One recent response to these types of obstacles has been a change in the locus and nature of globalised environmental regulation and protection, through a slow shift from state-centric to civil society-centred institutions. For much of the post-World War Two period, such international regulation as there was came about as the result of intergovernmental negotiations, and led to the signing and ratification of agreements by individual states. While non-state actors were permitted access to and influence in a number of intergovernmental institutions, for the most part they were quite restricted in terms of what they could say or do in such fora.

The past 15 years or so have seen considerable changes in this pattern. Now, not only are 'nongovernmental actors' achieving growing representation in international settings, they are also establishing a growing number of semi-public and private 'regimes' that are meant to fulfil a regulatory function that the UN system is unable or refuses to address. Moreover, many of these nongovernmental groups and organisations are taking upon themselves the functional responsibilities for seeing that certain regulations – both national and international – are adhered to by both public and private actors *and*, in some instances, fulfil the regulatory requirements themselves.

Thus, we are witness to a growing number of what might be called 'nongovernmental regimes' operating at the global level. These arrangements, which run the gamut from public to private, represent attempts to offer regulation of environmental externalities to those actors that seek to have their activities certified as 'environmentally-friendly.' In addition, there is also a very large number of organisations and groups active at the local level, engaged in environmental protection, conservation and restoration. Often, these groups are linked to each other through various networks of communication, knowledge and action (Lipschutz, 1996b, ch. 3).

The growth of such organisations, institutions and practices at and across *all* levels of analysis, from the local to the global is quite striking (see below; also see Leatherman, Pagnucco & Smith, 1994). What this seems to suggest is an emerging system of *global governance*, in which institutionalised regulatory arrangements are being complemented by less formalised norms, rules and procedures that pattern behaviour without the presence of written constitutions or material power.[5] One arena in which this diffusion of institutions and authority is especially visible is in terms of sustainable forestry, as we shall see below.

III. LOCALISING FUNCTIONS AND GLOBALISING THEM

Fragmegration is thus best understood as an integral part of a dialectic of globalisation. The state remains the primary actor in *international relations*, but the *jurisdictional authority* long monopolised by states is being spread throughout an emergent, multi-level and, for the moment, very diffuse system of globalising governance. Within this system, 'local' management, knowledge and rule are as important to co-ordination within and among local, national and global political 'hierarchies', regions and countries as the international management manifested in traditional regimes and international organisations. The two apparently contradictory tendencies of integration and fragmentation actually involve the transfer of *functional responsibility and authority* downward to the regional and local levels as well as upward to the global level. All of this is taking place, moreover, with the full connivance of national governments, aided and abetted by a wide variety of other institutions and actors.

What will the resulting political arrangements look like? Some have suggested that such social innovation constitutes a 'new mediaevalism'; others have proposed as organising principles 'heteronomy' or 'heterarchy'.[6] In discussing the first of these three concepts, Ole Wæver argues that

> For some four centuries, political space was organised through the principle of territorially defined units with exclusive rights inside, and a special kind of relations on the outside: International relations, foreign policy, without any superior authority. There is no longer one level that is clearly *the* most important to refer to but, rather, a set of overlapping authorities (Wæver, 1995, p. 59; emphasis in original).

What is critical in Wæver's point is not territorial control, but *authority* – In the sense of the ability to get things done because of one's legitimacy, as opposed to one's ability to apply force or coercion. Such distributed authority arrangements are clearly seen with respect to environmental issues and policy (see below).

In this globalised 'heteronomy,' authority is being distributed among many centres of political action, often on the basis of regulation focused on specific issue areas and problems, rather than exercised within nationally-bounded territories. This authority is developing, in part, because specific collectivities are taking over responsibilities no one else wants and, in part, because of expertise acquired through global networks of knowledge and practice (Lipschutz, 1996b, ch. 3,8). The redistribution of authority generates a form of globalised functional differentiation rather than world federalism, inasmuch as different authorities are being called on to deal with specific applied problems – toxic wastes, forest preservation, marshland protection – rather than a broad range of generalised ones, as the state does now. Because these problems are embedded within a global economic system, such local functionalism also reaches beyond localities into and through that global system, as groups communicate with and draw on the experience and expertise of similar collectivities elsewhere throughout the world. Note that this version of functionalism is not the same as that offered by Mitrany (1966) or the neofunctionalism of Haas (1964). Whereas those theories envisioned political *integration* as the outcome of international functional co-ordination, it is much more likely that this type of functionalism will operate at multiple levels without necessarily fostering political integration.

Amidst 'fragmegration', what constitutes the new loci of functional authority? There is growing evidence that the emergence of civil society organisations is one response to the lacuna of world government or international regulation (Lipschutz, 1996b). One of the more conspicuous phenomena in world politics over the past two or three decades has been the explosive growth in non-corporate, nongovernmental organisations, acting across a broad range of issue areas, at the local, national and global levels. Although no one is sure of the number, there may be tens of thousands (or more) of such organisations around the world, engaged with matters of development policy, environment, human rights, feminism, gender and culture, among others.

Reflecting the efflorescence of such activities, research into transnational social movements, transnational actor networks (TANs), nongovernmental organisations (NGOs), global networks and coalitions, and global governance structures has become a popular and profitable endeavour. Virtually all of the relevant works take note of extensive patterns of transnational interaction that suggest a fundamental change in world politics (see, e.g., Princen & Finger, 1994; Wapner, 1996; Bollier, 1997; Hirst, 1997; Keck & Sikkink, 1998; Korten, 1998). I characterise the entirety of these groups and their activities as 'global civil society'. These civil associations, understood in the broadest sense, fulfil multiple roles and, although their internal balance between altruism and selfishness can vary greatly, some of both can be found in all of them. Indeed, many associations that fall into this category go beyond collective self-interest to active pursuit of the collective political good (Lipschutz, 1996a; Lipschutz, 1996b, ch. 4-6).

What is important for the present analysis is that political civil society and the state are neither independent nor autonomous of each other. A state, whether liberal or not, relies on some version of civil society for its legitimacy. Conversely, a civil society cannot thrive without the legitimacy bestowed on it by the state, whether or not its government is democratic. One example of this sometimes contradictory relationship was visible in pre-1998 Indonesia (and continues today), where the authoritarian New Order state of President Suharto tolerated, and sometimes encouraged, a civil society of as many as 11,000 nongovernmental organisations (Laber, 1997, p. 43). Many of these groups occupied a kind of uncertain middle ground, supporting the state in some instances, opposing it in others. Similarly, at times, the state found it necessary to rely on these groups for certain functional needs as well as support. Two environmental umbrella organisations, WAHLI (the Indonesian Environmental Forum) and SKEPHI (Indonesian Network for Forest Conservation) were especially influential in challenging the environmental cronyism of the Suharto regime (Lipschutz, 1996b, ch. 5).

Another example – this time transnational – is the Climate Action Network (CAN),[7] a global alliance of regional coalitions made up of national and local environmental organisations and individuals. The members of CAN are engaged in a continuous and reciprocal exchange of knowledge and practice, some of it universal, some of it contingent and contextual (Lipschutz, 1996b, ch. 3). Members and organisations participate in local educational activities, regional and national lobbying, and international negotiations, such as those dealing with the UN Framework Convention on Climate Change. They act as technical and political advisors to governments and their agencies in some settings, and as adversaries of governments in others. They become involved in management at the local, regional, national and global levels, both as Network and as individuals and groups in specific locales. CAN is only one such network of global civil society; there a many more, some of them subnational, some among a few countries, others with global reach.

The diffusion of formerly state-centred regulatory responsibility and authority can be observed in the sustainable forestry sector. In the area of forestry practices, regulation through conventional international regimes has been limited, although several efforts, such as the Forestry Principles offered at UNCED in 1992, remain on the agenda, while others, such as the Tropical Forests Action Plan, have proved less than successful. The recent trend in this area appears to be away from regulation, per se, and toward certification of both national and private practices through what is called 'eco-labeling'. An eco-label is a claim placed on a product that is intended to enhance the item's social or market value by conveying its environmentally advantageous components (Markandya, 1997).

While there is a growing number of organisations – public, semi-public, and private – offering such labeling, the Forest Stewardship Council (FSC) illustrates one typical case of global regulation through the activities of global civil society. The FSC was launched in 1992 at a meeting in Washington, DC by a loose alliance of environmental groups, led by the World Wildlife Fund. An interim board was elected, a mission statement adopted, and draft Principles and Criteria for Forest Management initiated shortly thereafter. From its base in Oaxaca, Mexico, the Forest Stewardship Council has become an internationally recognised organisation with nearly 200 members in 50 countries. Structurally, the FSC is a membership organisation comprised of three equally weighted chambers: environmental, social and economic; membership within each chamber is also equally weighted between North and South.

Each chamber represents 33 per cent of the vote at Annual Meetings, and the Board of Directors has rotating members reflecting these interests. With international governmental processes in apparent stalemate, the FSC is seen increasingly by many as an institution able to fill a critical niche towards achieving sustainable forest management where governments cannot or will not.

Yet another form of environmental functionalism can be seen in two examples of the 'local-global' phenomenon mediated through networks of knowledge and practice, the Global Rivers Environmental Education Network (GREEN), which has projects in 136 countries, and the River Watch Network (RWN), which is based in Vermont.

> GREEN seeks to improve the quality of watershed and rivers, and thereby the lives of people. GREEN uses watersheds as a unifying theme to link people within and between watersheds. ... Each watershed project is unique, and how it develops depends upon the goals and situation of the local community ... As they share cultural perspectives, students, teachers, citizens and professionals from diverse parts of the world are linked by a common bond of interest in and concern for water quality issues (quoted in Lipschutz, 1996b, p. 150).

RWN is more focused on the linkages between technological and scientific competence and political action, without much reference to larger goals. As RWN's materials put it, 'We can help you clean up your river'.

> River pollution ... is generated by all of us and its solution requires active citizen participation. Federal, state and local governments are frequently unable to tackle these water quality problems because their resources for river monitoring are severely limited. ... Gathering and interpreting scientifically credible water quality data underlies every River Watch effort. ... RWN will never just send you a kit with a page of instructions for water sampling. ... Each River Watch

program is individually designed to meet the particular needs of its community and the conditions of its river. ... RWN staff are river experts *and* community organisers. (quoted in Lipschutz, 1996b p. 150; emphasis in original).

Both GREEN and RWN are only part of a growing worldwide effort to protect and restore river and stream watersheds. Sometimes, such groups are affiliated with organisations such as GREEN and RWN, at other times they are entirely local, their establishment having been inspired by similar groups and practices elsewhere.[8]

In more general terms, we can identify four modes in which global civil society is organised and engaged in practices intended to change the global landscape of environmental regulation. These include: (1) ecosystem management and restoration; (2) fostering of localised environment/development projects; (3) environmental education; and (4) participation in national and transnational networks and alliances (Lipschutz, 1966b, ch.2-3).

Ecosystem management and restoration

There are large (and growing) numbers of quasi- and nongovernmental groups and organisations engaged in the management and restoration of environmental resources, often on an ecosystem basis. For example, individuals working on a small-scale watershed restoration project in the Sierra Nevada foothills of California or the river bottoms of Hungary might not think of themselves as being linked into global networks. But they often receive visitors from other parts of the United States and the world, who come to study the project as a model for restoring other watersheds. Those projects, in turn, inform others, and so on. There are also organisations with a wider scope that help to support, technically and financially, similar linked efforts in the United States and other countries (Lipschutz, 1996b, ch.5). Individually, such projects might not seem very significant, but in aggregate they are, for two reasons. First, each stands as a form of social organisation that can be studied and reproduced elsewhere, from both the technical and social perspective. Second, each also fulfils an *educational* function and draws local community members into the regime reconstruction project. Thus, practice begets knowledge and knowledge begets practice.

Local environment/development projects

Throughout the world, in industrialised countries as well as developing ones, there are burgeoning numbers of small, locally-oriented organisations engaged in the provision of a vast range of services to marginal and neglected populations, that incorporate environmental concerns. Often, these projects are initiated in a wholly-local fashion in a manner that is fairly autonomous of the overarching state. For example, Brazil is the site of growing numbers of urban-based groups, with roots in the Christian base communities that emerged during the dictatorship, who now seek to link environmental concerns with problems of urban pollution and economic justice (Jamie Anderson, 1994). Many of these groups also draw on technical and financial assistance through global networks and transnational alliances established with other organisations in industrialised countries and even with the agencies of developed country governments.

Environmental education

A third approach depends on locally based activists who undertake education, demonstrations, and proselytising on behalf of environmental protection as well as the conservation of specific resources. Some of these groups are engaged in an effort to revise the constitutive basis for relationships between human society and nature; others are less ambitious in their goals. Although environmental activists have, traditionally, been urban and suburban, they are appearing in increasing numbers in rural areas where many of the resources of greatest concern are actually located. Here, they are having an impact far beyond their rather limited numbers (Lipschutz, 1996b, ch.4).

National and transnational networks and alliances

The fourth element is based on networks and alliances, national, transnational and global, linked together by what their members see as common strategies and goals. All of these networks exist under the over-arching rubric of a general environmental ethic – what might be called an 'operating system' – although the specific form of relations through the network and structure of the actors at the ends and nodes of the network vary a great deal. Some of these networks are quite deliberately contra-state, others are oriented toward state reform; a few are both. Some networks simply ignore the state altogether.

For example, WAHLI, mentioned above, is a network of some 300 environment/ development groups in Indonesia that sometimes works with state agencies and, at other times, directly opposes them (Lipschutz, 1996b, ch. 6). Greenpeace is something of a global network in itself, with both contra-state and state-reforming tendencies (Wapner, 1996), while a number of observers even tend to view Greenpeace as a purveyor of environmental 'imperialism' (Seager, 1993, ch. 4). The Asian Pacific People's Environmental Network, based in Penang, Malaysia, includes both urban and rural organisations, and operates at the international and regional levels. Indigenous peoples groups and coalitions are growing in number and influence (Wilmer, 1993). And, as noted earlier, the Climate Action Network has branches around the world.

The emergence of institutions of environmental functionalism can be understood as a consequence of social *innovation*, of the generation of new scientific-technical and social knowledge(s) required to address different types of contemporary issues and problems, and of the reorganisation of social formations and practices, all arising from globalisation.[9] Inasmuch as there is too much scientific and social knowledge for any one actor, whether individual or collective, to assimilate, it becomes necessary to establish knowledge-based alliances and coalitions whose organisational logic is only partly based on space or, for that matter, hierarchy. 'Local' knowledge is spatially-situated while 'organisational' knowledge – how to put knowledge together and use it – is spaceless (but possibly time dependent). Combined together, the two become instrumental to rapid technical and social innovation at a scale, and in a manner, that leads to broad social change and reorganisation.

Acquisition of functional knowledge and practices leads, in turn, to new forms and venues of authority (see Rosenau, 1990, ch. 14), in that only those with access to such capabilities can act successfully. Successful action, based on knowledge and practice, generates legitimacy and authority. Completing the circle, those organisations with such legitimacy and authority are then sought out for their management expertise when other circumstances appear, similar to those that first led to the group's launch. Some of this expertise is scientific in the traditional sense, some of it is contextual and contingent in the sense of being 'indigenous knowledge' (Lipschutz, 1996b, ch. 4-6). The locus of such functionalist governance varies, but it is more likely to be regional or local – in the lab, the research group, the neighborhood, the watershed – than in some abstract and undifferentiated global realm. As the late Richard Gordon put it,

> Regions and networks ... constitute interdependent poles within the new spatial mosaic of global innovation. Globalisation in this context involves not the leavening impact of universal processes but, on the contrary, the calculated synthesis of cultural diversity in the form of differentiated regional innovation logics and capabilities. ... *The effectiveness of local resources and the ability to achieve genuine forms of cooperation with global networks must be developed from within the region itself.* (Gordon, 1995, pp. 196, 199; emphasis added).

Such functionalist localisation also helps to illuminate on part of the puzzle of fragmegration: Lines must be drawn somewhere, whether by reference to Nature, power, authority or knowledge. From a constructivist perspective, such lines may be as 'fictional' as those which currently separate one country, or one county, from another. Still, they are unlikely to be wholly disconnected from the material world, inasmuch as they will have to map onto already existing patterns and structures of social and economic activity (see Lipschutz 1999a).

IV. THE FUTURE OF GLOBALISATION AND GLOBAL ENVIRONMENTAL GOVERNANCE

The emergence of global civil society as a source of activism at this juncture suggests, in part, a reorganisation of global politics away from state-centric institutions to arrangements in which authority is much more diffuse and sometimes open for the taking (Rosenau, 1997). With a few exceptions (Wapner, 1996; Hirst, 1997; Krut, 1997; Korten, 1998), most of the current literature on global civil society and similar forms of political activity does not recognise this change, speaking instead of the decline in state sovereignty and the growing porosity of national borders (Gill, 1994). Similarly, much of the comparative literature focused on national activities of such groups recognises the existence of cross-border alliances and coalitions, such as the Climate Action Network and the Asia Pacific People's Environmental Network, but it tends to view them through comparative politics or within the confines of national politics (Bollier, 1997). The ways in which globalisation and integration are linking the local, national and transnational in political practice, while acknowledged, is rarely problematised or examined closely (see Lipschutz, 1996b). But what the emergence of global civil society does indicate is that political community – even in a federal state – is not restricted to discrete levels of government, and that governance is more than just governments.

Why should political reorganisation be taking place through global civil society? Is this really something new, or has the phenomenon been seen before (Murphy, 1994)? My explanation rests on certain notions about global political economy since 1945. One of the consequences of the 'information revolution', rooted in the discovery and application of new knowledge through a Fordist version of scientific research on a large scale during and after World War Two, was a vast expansion in the supply of higher education throughout the world (Lipschutz, 1997a). To be sure, the enormous growth in sources and availability of information and the increase in modes of transnational communication were both empowering, as Rosenau (1990) has argued, but for most of the 1950s and well into the 1960s, these 'powerful people' were absorbed rather easily by state and economy, where they supported existing structures of authority. What changed?

In many countries, the expansion of higher education led to a large increase in the supply of college graduates, a supply that began to exceed demand. Beyond a certain point, state control of the economic and political systems made expansion of employment opportunities problematic (a problem evident today across Europe). After certain political failures, such as the Vietnam War, the legitimacy and authority of the state was seriously weakened. In the United States, growing numbers of educated cadres struck out on their own, establishing the plethora of consultancies, think-tanks and other such institutions – many of them with an environmental bent – that are so familiar to us today. In this way, these cadres hoped to do better than the state by disseminating knowledge and practice throughout the society and the world (see Lipschutz, 1997a; Lipschutz, 1999b, ch. 2).

The growth in the numbers of educated people has been globally important in instrumental terms, but to explain the longer-term significance of this phenomenon we need to look not only at material factors or institutions, but also to the progenitors of ideas and action. For this purpose, it is useful to seek historical parallels. Michael Mann (1993) draws on the writings of Antonio Gramsci to explain the emergence of national states in Europe and North America during the 1700s and 1800s, and the economic, political and social revolutions and changes that took place throughout the 'long' 19th century. Put briefly, Mann sees the rise of what Gramsci called 'organic intellectuals' as central to the transition from royal to popular sovereignty and the creation of the modern state. These organic intellectuals filled a discursive role in a gradual process of social change by developing and articulating the ideas and practices that animated the political and social upheavals of those times. Mann observes that, while material interests and needs were always central to popular mobilisation, emotional and ideational incentives were as important, if not more so.

Beyond this, the ideas and arguments propounded by the organic intellectuals were framed in terms of 'progress', promising a better future through political, economic and social reorganisation. The ideologies of nationalism, liberalism, socialism and other -isms that reified the strong state were created by these organic intellectuals. Without the communication and putting into practice of their arguments, the 19th and 20th centuries might have been much quieter times. As it was, the centralised nation-states that dominated world politics for the past century were, for better or worse, legitimated by the ideas of these intellectuals, if not constructed by them. As Mann has put it

> Capitalism and discursive literacy media were the dual faces of a civil society diffusing throughout eighteenth-century European civilisation. They were not reducible to each other, although they were entwined. ... Nor were they more than partly caged by dominant classes, churches, military elites, and states, although they were variably encouraged and structured by them. Thus, they were partly transnational and interstitial to other power organisations. ... Civil societies were always entwined with states – and they became more so during the long nineteenth century (1993, p. 42).

Mann's arguments have an interesting contemporary resonance, and they force us to ask: Where are today's organic intellectuals? And what are they up to?

I suggest that they are to be found thinking and acting through the groups, movements and institutions of global civil society. The participants in these collectivities constitute a cadre of transnational intellectuals – not necessarily because they travel, but because they engage in transnational 'discursive literacy' – who are filling a role in the development of systems of global governance similar to those that developed in national settings in centuries past. These organic intellectuals are also filling what is, so far, a rudimentary representational role at the global level through the institutions and organisations of global civil society (which are not, so far, very representative). Indeed, just as modern democracies came into being in a symbiotic relationship with their domestic civil societies, so is this emergent system of global governance coming into being in concert with a 'global civil society.'

Again, the environmental issue area provides one of the clearest and most visible examples of the activities of such organic intellectuals. Some are engaged in the writing of popular books and the production of television programmes on a range of environmental problems. Many regularly attend the environmental gatherings, workshops and conferences that are held throughout the world on an almost daily basis. Still others are engaged in providing advice to politicians, policy makers, bureaucrats and legislative representatives, through briefings, books, reports, studies, op-ed pieces, news conferences and so on. Amidst all of this, it is important not to forget the role of educational institutions. Since the early 1960s, these have educated a whole generation of college graduates in environmental science and environmental studies. Some graduates now teach ecological principles to first graders, others have changed the structure and belief systems of long-established bureaucracies and corporations. A few are even to be found in the upper reaches of national governments. Through their activities, basic information about the environmental crisis and proposals for problem-solving action are being disseminated widely. As people become more familiar with these ideas, and begin to act on them, world politics will begin to change.

There is a strong intersubjective element at work here: analysis informs action and vice versa. Today's organic intellectuals engage in the examination of transnational phenomena and articulate their significance (see Keck & Sikkink, 1998), eschew determinism and offer alternative conceptualisations of how things might be done. They also transfer both knowledge and practice via national and transnational coalitions, alliances and communications, and create the organisations and institutions that propagate these notions and carry them to various levels of government and governance. In this way, the architecture of a new global politics is emerging in the interstices of the current world system, just as the state system and capitalism grew out of feudalism centuries ago (Mann, 1993, ch. 2).

V. IN CONCLUSION

Is there hope for the environment? Can the world's peoples be mobilised to act on its behalf, to take on the functional task of conserving, protecting and restoring their part of the world's environment, and to facilitate transnational collaboration through a variety of institutions that will help to bring to fruition a new form of global politics and also save the global environment? I have argued here that one possible answer and mode of action can be found in the environmental politics and policy making practiced by global civil society. To return to one of the questions posed at the outset of this paper, if anything *is* to be done, it will be done as a result of social change, fostered through global civil society and the people who make it up.

NOTES

1. Portions and versions of this chapter have appeared in other places, including Lipschutz (1998b, 1997b, 1996a and 1996b).

2. There are some who might dispute this point; see, e.g., the work associated with the Marshall Institute in Washington, D.C.

3. By this definition, therefore, civil society includes social movements, various kinds of public interest groups, and corporations (although I am not explicitly discussing the last here), all of which do engage in politics of one sort or another. The state-civil society distinction is, sometimes, difficult to ascertain, as in the case of the World Wildlife Fund/Worldwide Fund for Animals and other similar organisations, which subcontract with state agencies.

4. Rosenau (1990) has taken these contrary tendencies into account by theorising 'sovereignty-bound' and 'sovereignty-free' actors. This, I think, does not capture the entire dynamic, in that some of the actors in the latter category would dearly love to move into the former.

5. This point is a heavily disputed one: To wit, is the international system so undersocialised as to make institutions only weakly-constraining on behaviour, as Stephen Krasner (1993) might argue, or are the fetters of institutionalised practices sufficiently strong to modify behaviour away from chaos and even anarchy, as Nicholas Onuf (1989) might put it?

6. The best-known discussion of the 'new mediaevalism' is to be found in Bull (1997, pp. 254-55, 264-76, 285-86, 291-94). The notion of 'heteronomy' is found, among other places, in Ruggie (1983, p. 274, n. 30). A heterarchy is a system of functionally-differentiated and overlapping sovereign and semi-sovereign authorities, while a society is composed of individual entities embedded in webs of institutional relationships. The term 'heterarchy' comes from Bartlett & Ghoshal (1990), quoted in Gordon (1995, p. 181).

7. For information about the Climate Action Network, see http://www.igc.org/climate/Eco.html.

8. A more detailed discussion of this argument can be found in Roger A. Coate, Chadwick F. Alger & Ronnie D. Lipschutz, 'The United Nations and Civil Society: Creative Partnerships for Sustainable Development', *Alternatives* 21, 1 (Jan.-Mar. 1996):93-122.

9. The following paragraphs are based on Gordon (1995). He argues for the existence of three 'logics' of world-economic organisation: internationalisation; multi-/transnationalisation; and globalisation. The last is 'heterarchical' and non-market and, as he puts it, involves 'valorisation of localised techno-economic capabilities and socio-institutional frameworks... [with] mutual reciprocity between regional innovation systems and global networks' ('Concurrent Processes of World-Economic Integration: A Preliminary Typology,' handout in colloquium, Nov. 30, 1994, UC-Santa Cruz).

REFERENCES

J.J. Anderson, 'Social Movements and Environmental Politics in Urban Brazil', Paper prepared for delivery at the 1994 Annual Meeting of the American Political Science Association, New York Sept. 1-4, 1994.

B. Barber, *Jihad vs. McWorld* (New York: Times Books, 1995).

C. Bartlett & S. Ghoshal, 'Managing Innovation in the Transnational Corporation,' in C.Y. Doz & G. Hedlund (eds), *Managing the Global Firm* (London: Routledge, 1990) pp. 215-55.

J. Bhagwati, 'Trade and the Environment: The False Conflict?' in D. Zaelke, P. Orbuch & R.F. Houseman (eds), *Trade and the Environment: Law, Economics, and Policy* (Washington, D.C.: Island Press, 1993) pp. 159-90.

D. Bollier, *Beyond Bureaucracy: New Roles for Government, Civil Society and the Private Sector* (Washington, DC: Aspen Institute, 1997).

H. Bull, *The Anarchical Society: A Study of Order in World Politics* (New York: Columbia University Press, 1977).

M. Castells, *The Information Age* (Malden, Mass: Blackwell, 1996, 1997, 1998; 3 vol.).

E.O. Czempiel, 'Governance and Democratization,' in James N. Rosenau & Ernst-Otto Czempiel (eds.), *Governance without Government: Order and Change in World Politics* (Cambridge: Cambridge University Press, 1992) pp. 250-71.

S. Gill, 'Structural change and the global political economy: Globalizing elites and the emerging world order', in Yoshikazu Sakamoto (ed.), *Global transformation: Challenges to the state system* (Tokyo: United Nations University Press, 1994) pp. 169-99.

R. Gordon, 'Globalization, New Production Systems and the Spatial Division of Labor', in W. Litek and T. Charles (eds), *The Division of Labor: Emerging Forms of World Organization in International Perspective* (Berlin: Walter de Gruyter, 1995) pp. 167-207.

E.B. Haas, *Beyond the Nation-State* (Stanford: Stanford University Press, 1964).

P. Haas, R.O. Keohane & M.A. Levy (eds), *Institutions for the Earth: Sources of Effective International Environmental Protection* (Cambridge, Mass: MIT Press, 1993).

P. Hirst, *From Statism to Pluralism: Democracy, Civil Society, and Global Politics* (London: UCL Press, 1997).

R. Kaplan, *The Ends of the Earth: A Journey at the Dawn of the 21st Century* (New York: Random House, 1996).

M. Keck and K. Sikkink, *Activists Across Borders: Advocacy Networks in International Politics* (Ithaca, NY: Cornell University Press, 1998).

R.O. Keohane & M.A. Levy (eds), *Institutions for Environmental Aid: Pitfalls and Promise* (Cambridge, MA: MIT Press, 1996).

D. Korten, *Globalizing Civil Society: Reclaiming our right to power* (New York: Seven Stories Press, 1998).

S. D. Krasner, 'Westphalia and All That', in: Judith Goldstein & Robert O. Keohane (eds), *Ideas and Foreign Policy* (Ithaca: Cornell University Press, 1993).

R. Krut, *Globalization and Civil Society: NGO influence in international decision-making* (Geneva: UN Research Institute for Social Development, 1997).

J. Laber, 'Smoldering Indonesia', *New York Review of Books* 44, 1 (Jan. 9, 1997): 40-45.

S. Lash, B. Szerszynski & B. Wynne (eds.), *Risk, Environment & Modernity: Towards a New Ecology* (London: Sage, 1996).

J. Leatherman, R. Pagnucco and J. Smith, 'International Institutions and Transnational Social Movement Organizations: Transforming Sovereignty, Anarchy, and Global Governance', Kroc Institute for International Peace Studies, Univ. of Notre Dame, August 1994, Working Paper 5.

R.D. Lipschutz, 'Members Only? Citizenship and Civic Virtue in a Time of Globalization,' *International Politics* 36, 2 (June 1999a).

R.D. Lipschutz, *After Authority: War, Peace and Global Politics in the 21st Century* (Albany: SUNY Press, 1999b).

R.D. Lipschutz, 'Seeking a State of One's Own', in: B. Crawford & R.D. Lipschutz, *The Myth of `Ethnic Conflict': Politics, Economics and `Cultural' Violence* (Berkeley: International & Area Studies Press, UC-Berkeley, 1998a).

R.D. Lipschutz, 'The Nature of Sovereignty and the Sovereignty of Nature: Problematizing the Boundaries between Self, Society, State and System,' in K.D. Litfin (ed.), *The Greening of Sovereignty in World Politics* (Cambridge, MA: MIT Press, 1998b).

R.D. Lipschutz, 'The Great Transformation Revisited', *Brown Journal of International Affairs* 4, 1 (Winter/Spring 1997a) 299-318.

R.D. Lipschutz, 'From Place to Planet: Local Knowledge and Global Environmental Governance', *Global Governance* 3, 1 (Jan.-Apr. 1997b) pp. 83-102.

R.D. Lipschutz, 'Reconstructing World Politics: The Emergence of Global Civil Society', in: Jeremy Larkins & Rick Fawn (eds), *International Society after the Cold War* (London: Macmillan, 1996a) pp. 101-31.

R.D. Lipschutz, with J. Mayer, *Global Civil Society and Global Environmental Governance* (Albany, NY: SUNY Press, 1996b).

M. Mann, *The Sources of Social Power: The rise of classes and nation-states, 1760-1914* (Cambridge: Cambridge University Press, 1993, vol. II).

A. Markandya, 'Eco-labeling: An Introduction and Review,' *Eco-labeling and International Trade* (New York: St. Martin's, 1997).

D. Mitrany, *A Working Peace System* (Chicago: Quadrangle Books, 1966).

C. Murphy, *International Organization and Industrial Change: Global Governance since 1850* (New York: Oxford University Press, 1994).

K. Ohmae, *The End of the Nation State* (New York: Free Press, 1995).

N.G. Onuf, *World of Our Making: Rules and Rule in Social Theory and International Relations* (Columbia, SC: University of South Carolina Press, 1989).

T. Princen & M. Finger (eds), *Environmental NGOs in World Politics* (London: Routledge, 1994).

R. Putnam, 'Diplomacy and Domestic Politics: The Logic of Two Level Games', *International Organization* 42, 3 (Summer 1988):427-60.

M. Redclift & T. Benton (eds), *Social Theory and the Environment* (London: Routledge, 1994).

J.N. Rosenau, *Along the Domestic-Foreign Frontier: Exploring Governance in a Turbulent World* (Cambridge: Cambridge University Press, 1997).

J.N. Rosenau, *Turbulence in World Politics* (Princeton, NJ: Princeton University Press, 1990).

J.G. Ruggie, 'Continuity and Transformation in the World Polity: Toward a Neorealist Synthesis', *World Politics* 35, 2 (Jan. 1983) 261-85.

J. Seager, *Earth Follies: Coming to Feminist Terms with the Global Environmental Crisis* (New York: Routledge, 1993).

S. Strange, *The Retreat of the State* (Cambridge: Cambridge University Press, 1996).

S.K. Vogel, *Freer Markets, More Rules: Regulatory Reform in Advanced Industrial Countries* (Ithaca: Cornell University Press, 1996).

O. Wæver, 'Securitization and Desecuritization', in Ronnie D. Lipschutz (ed.), *On Security* (New York: Columbia University Press, 1995) pp. 46-86.

P. Wapner, *Environmental Activism and World Civic Politics* (Albany: SUNY Press, 1996).

F. Wilmer, *The Indigenous Voice in World Politics* (Newbury Park: Sage, 1993).

P. Woodall, 'The World Economy: Who's in the driving seat?' *The Economist*, Oct. 7, 1995, special insert.

O.R. Young, *International Governance: Protecting the Environment in a Stateless Society* (Ithaca: Cornell University Press, 1994).

D. Zaelke, P. Orbuch & R.F. Houseman (eds.), *Trade and the Environment: Law, Economics, and Policy* (Washington, DC: Island Press, 1993).

Index